PRAISE FO

James, John W.
When children
grieve

AUG 0 8 2005

"In my t̶... never seen a more u̶... book tran- scends tr̶... n practical, everyday language, the authors present an action-oriented program that provides concrete suggestions of things to say and do with a child to help acknowledge the loss and to 'complete' the process of grieving in the healthiest of ways. Although this book positions adults as the child's 'emotional leader' in the grief process, all adults can benefit greatly by applying these principles to themselves as well."

—Edwin S. Cox, Ph.D.,
president emeritus, Phillips Graduate Institute

"Once in a generation, a book comes along that alters the way society views a topic. We teach our children to read, write, and do math so their lives will be successful. While gains are important, loss is unavoidable. Sadly, we do not always teach our children useful skills to deal with the feelings caused by losses, large and small—mainly because we do not know how. *When Children Grieve* is an essential primer for parents and others who interact with children on a regular basis."

—Bernard McGrane, Ph.D.,
professor of sociology, Chapman University
and University of California at Irvine

DISCARD

The first of its kind, *When Children Grieve* teaches parents how to initiate in their children lifelong, healthy response patterns to grief and to empower them with effective methods for dealing with loss.

There are many life experiences that can produce feelings of grief in a child—everything from the death of a relative to a divorce to everyday experiences such as moving to a new neighborhood or losing a prized possession. No matter what the reason or the degree of severity, if a child you love is grieving, the guidelines examined in this thoughtful book can make a difference. For example:

- Listen with your heart, not your head. Allow all emotions to be expressed without judgment, criticism, or analysis.
- Recognize that grief is emotional, not intellectual. Avoid the trap of asking your child what is wrong, because he or she will automatically say "nothing."
- Adults—you go first. Telling the truth about your own grief will make your child feel safe about opening up, too.
- Remember that each of your children is unique and each has a unique relationship to the loss event.
- Be patient. Don't force them to talk.
- Never say "Don't feel sad" or "Don't feel scared." Sadness and fear, the two most normal feelings attached to loss of any kind, are essential to being human.

No matter what the journey involves, witnessing the return of hope to a child's life is one of the greatest gifts you can give to any child you love.

About the Authors

© KATHY HUTCHINS

JOHN W. JAMES was born in Danville, Illinois. He was thrust, unwillingly, into the arena of grief and recovery when his three-day-old son died in 1977. John lives in Los Angeles with his Emmy Award–winning wife, Jess Walton—the evil Jill Abbott on *The Young and the Restless*—and spends most of his free time with his daughter Allison, twenty-five, and son Cole, twenty.

© JESSICA SCHAEFFER

RUSSELL P. FRIEDMAN was born in Port Chester, New York. He arrived at the Grief Recovery Institute in 1986, following a second divorce and a major financial disaster. He started as a volunteer and stayed and stayed and stayed. The apple of his eye, his daughter Kelly, lives in Los Angeles. Russell lives in Sherman Oaks with Alice Borden and their dog, Max.

© KAREN HUTER

DR. LESLIE LANDON MATTHEWS was born in Los Angeles. She attended a Grief Recovery Personal Workshop after the much-publicized death of her father, Michael Landon. She earned her doctorate in the field of psychology with a focus on children and grief. She lives in southern California with her husband, Brian, and their glorious children.

JOHN W. JAMES AND RUSSELL P. FRIEDMAN have been working with grievers for more than twenty years. They have served as consultants to thousands of bereavement professionals and provide grief recovery seminars and certification programs throughout the United States and Canada.

ALSO BY JOHN W. JAMES AND RUSSELL FRIEDMAN

*The Grief Recovery Handbook: The Action Program for Moving Beyond
Death, Divorce, and Other Losses*

WHEN CHILDREN GRIEVE

For Adults to Help Children
Deal with Death, Divorce,
Pet Loss, Moving, and Other Losses

JOHN W. JAMES AND RUSSELL FRIEDMAN
WITH Dr. Leslie Landon Matthews

Quill

An Imprint of HarperCollins *Publishers*

A hardcover edition of this book was published in 2001 by HarperCollins Publishers.

WHEN CHILDREN GRIEVE. Copyright© 2001 by John W. James and Russell Friedman. All rights reserved. Printed in the United States of America. No part of this book may be used or reproduced in any manner whatsoever without written permission except in the case of brief quotations embodied in critical articles and reviews. For information address HarperCollins Publishers Inc., 10 East 53rd Street, New York, NY 10022.

HarperCollins books may be purchased for educational, business, or sales promotional use. For information please write: Special Markets Department, HarperCollins Publishers Inc., 10 East 53rd Street, New York, NY 10022.

First Quill edition published 2002.

Designed by Jackie McKee

The Library of Congress has catalogued the hardcover edition as follows:

James, John W.

    When children grieve: for adults to help children deal with death, divorce, pet loss, moving, and other losses/ John W. James and Russell Friedman with Leslie Landon Matthews.—1st ed.

    p. cm.

  ISBN 0-06-019613-0

    1. Grief in children. 2. Loss (psychology) in children. 3. Bereavement in children. 4. Children and death—counseling of. I. Friedman, Russell. II. Matthews, Leslie Landon. III. Title.

BF732.G75J36 2001

155.9'3—dc21

ISBN 0-06-008429-4 (pbk.)

    05 06 ❖ RRD 10 9 8 7 6

At The Grief Recovery Institute our daily corporate pledge is to "never grow up."

So far, we think we're doing a pretty good job.

We dedicate this book to the child inside us all, big and small.

From John
In memory of Coach Paul Shebby,
who taught me self-discipline
Thanks, Coach

From Russell
This one's for you, Dad
And for my daughter, Kelly
with all my love

From Leslie
For my husband, Brian, and my dad, Michael
The men whom I will love forever

Contents

Foreword to the Quill Edition

In each of our lifetimes, there will be events that fall well outside the range of "normal," and for which we have little or no preparation or experience.

The events of September 11, 2001, are clearly so abnormal that most of us, adult or child, do not have a reference point to guide us in the emotional responses to and the aftermath of such an experience.

This book was written long before the tragic events of September 11 and makes no direct reference to it or any other such national or global occurrence.

Yet, what this book teaches about children's reactions to all loss events, and what it demonstrates to assist parents in guiding children, is practical and functional when applied to the kinds of feelings and fears caused by the "abnormal" incidents that generate an overwhelming amount of emotional energy in children.

Introduction

PUT *YOUR* OXYGEN MASK ON FIRST

At the beginning of every commercial airline flight, the cabin staff make a series of announcements about the safety procedures to follow in the unlikely event that an emergency should occur. Anyone who has flown will recall hearing what to do if the oxygen masks drop from the overhead compartment. The announcement sounds something like this: "If you are traveling with young children, fix the oxygen mask securely over your mouth and nose, *before* attempting to help your children." Each time you hear that, you automatically understand that if you can't breathe, you won't be able to help your children.

Since we are the safety attendants for this book, we are going to make the following announcement before this book taxies down the runway: "If you are traveling with children of any age, please affix the ideas in this book firmly over your head and your *heart, before* attempting to help your children."

Although we will not be passing through the cabin with refreshments, we will be giving you some incredibly valuable safety tips that can have a lifelong positive impact on your child's emotional well-being. Some of the ideas in this book will be new to you, and some will represent things you have known for a long time. The new ones may catch you off guard, and you may find yourself at odds with them at first, so with one last flight analogy, we alert you to: "Fasten your seat belts."

Who Are We?
And Why Have We Written This Book?

First, we are John W. James, who began helping grieving people twenty-three years ago, but not as a career choice. Like many people, John was drawn to this arena by an event so overwhelming that he did not know how to deal with it. That event was the death of his three-day-old son. John was so affected and limited by that loss, he realized that if he did not find a way to feel different or better, he did not think he could go on living in such pain for another forty or fifty years. John discovered a series of actions that helped him "complete" the pain in his heart, and allowed him to begin participating fully in life again. As John's friends noticed the changes in his outlook and attitudes, they began bringing others who were also struggling, just as he had been, to find a way to break the bondage of pain that seemed to rule their lives. Soon, John was spending all of his time helping grieving people, and his construction business fell by the wayside.

As John began to devote all of his waking hours to helping people deal with death, divorce, and other losses, he realized that he would never have enough time personally to help even a tiny fraction of the people whose lives had been devastated by tragic events. Inside this book is the detailed story that led to John writing the very first version of *The Grief Recovery Handbook*, which he self-published. One of the major reasons that John wrote that book was his realization that there were so many millions of grieving people and so little time to help them all. The book allowed people to learn about the actions they could use to help themselves, even though they did not have direct contact with John. Very soon after that self-published edition became available, HarperCollins (then called Harper & Row) published a new version of *The Grief Recovery Handbook, with the subtitle A Step-by-Step Program for Moving Beyond Loss*, which allowed John to upgrade and improve the ideas that he had been refining by working with grieving people.

Which brings us to coauthor Russell P. Friedman, who came to work with John in 1987. While Russell's life story is very different from John's, he also did not choose to be in this field. Russell arrived at The Grief Recovery Institute on the heels of a second divorce and a financial meltdown, both of which left him feeling emotionally crippled. He arrived at the institute not to help others, but to get help for himself.

In fact, Russell's first awareness of grief recovery came quite accidentally, when he was dragged by a friend to see John, who was presenting a short introductory lecture. Up until that day, Russell would not have known to use the word *grief* to explain the way he felt about his divorce or his financial dilemma. Like most people, Russell presumed that grief related only to death. And at that time, no one around Russell's life had died, so he didn't even understand why his friend had made him tag along to listen to John.

It was at that lecture that Russell heard some ideas that made him recognize that he indeed was a griever, and that the end of his marriage had meant the "death" of all the hopes, dreams, and expectations of going off into the sunset with that special someone. Russell contacted John the day after that lecture and volunteered to work at the institute. When John tells this story, he says, "Russell showed up one day thirteen years ago, and I haven't been able to get rid of him, so he is now my business partner, my coauthor, and my friend."

In 1998, HarperCollins published a revised edition of *The Grief Recovery Handbook: The Action Program for Moving Beyond Death, Divorce, and Other Losses*. Once more, John, this time with Russell's help, had the opportunity to refine and upgrade the actions that help people resume a productive place in the mainstream of their lives, even though they have experienced devastating losses.

And finally, we are Leslie Landon Matthews, Ph.D., who is affectionately known around here as "Fhud," which is how we pronounce Ph.D. She is also known as Dr. Mom, a fitting title to go along with the recent birth of her third child. Leslie arrived at the institute in an entirely unique manner, quite different from John and Russell.

Leslie's dad was Michael Landon, the actor known to nearly all from *Bonanza, Little House on the Prairie*, and *Highway to Heaven*. Michael Landon died in 1991 at the age of fifty-four. Any of us is likely to be devastated when a parent dies, but often the pain is made more intense when the parent is young and is robbed of what would be considered a normal life span. Children are left with unrealized expectations of what the future with their parents might have held. But a third dimension was added to Leslie's grief. Her dad's fame made it nearly impossible for Leslie and her family to deal privately with their emotions about his death. The onslaught of media attention spilled over even to the cemetery, making it impossible to find any solitude or comfort at her dad's grave.

Leslie attended a three-day Grief Recovery Personal Workshop to try to deal with the conflicting feelings caused by her father's death. She was somewhat surprised to discover that there was an almost equal intensity of pain attached to her parents' divorce several years earlier. That divorce had affected Leslie's relationship with her dad, and it was now time to complete some unfinished emotions that had been percolating for a long time.

When Leslie arrived at the institute, she was a practicing Marriage and Family Therapist (MFT), specializing in helping children. Yet all of her training had not prepared her for the mixture of emotions she experienced following her father's death. The positive emotional changes she perceived in herself after the personal workshop helped her see that The Grief Recovery Institute was delivering a level of assistance to grieving people that was different and far more effective than anything she had previously been aware of. Leslie realized that most of the children she was seeing in her capacity as a therapist were struggling with a variety of grief and loss issues. With the idea that she could apply the things she could learn from John and Russell to helping children, she attended a Grief Recovery Certification Training Program.

What Leslie didn't realize was that John and Russell also had a plan. They had been threatening to write a book about helping children deal with loss for a long time. But other urgent matters kept interfering. In spending time with Leslie, they recognized a unique combination of interests and personality that would make her an ideal candidate to tackle some of the issues that would be relevant to that kind of book. So they started promoting the idea of Leslie pursuing her Ph.D. on the topic of children and grief.

Poor Leslie, she was outnumbered; not only were John and Russell on her case, but her incredibly supportive husband, Brian, joined the team and urged her on. By this time, Leslie was a mommy, and a very dedicated one, so she had to find ways to be a full-time mommy and put in all the hours of research and writing to complete her Ph.D. She managed both, brilliantly.

John, Russell, and Leslie, each in turn, arrived at The Grief Recovery Institute to complete the unfinished emotional business contained in their relationships with people, some living and some dead. In addition, each of them discovered that they had a desire and willingness to help others whose lives had been affected by significant emotional losses of all kinds. That desire, coupled with accurate and effective information, helps them create safety so others can take new actions to complete the pain caused by loss.

We would like to encourage you to take the last paragraph to heart. We know that each and every one of you wants to help your children or the children in your care. Willingness, compassion, and love are wonderful motivators to help others. We support that idea, with just one little asterisk, and that is the fact that safety for your children is created by a combination of good intentions and *proper tools, skills and information.*

For this new beginning, from our hearts to yours and to your children.

John, Russell, and Leslie

P.S.: Grief❣Recovery®. Look for the Heart❣, which is our unique registered trademark. It will help you identify those people who have been certified and trained by The Grief Recovery Institute. Those who are authorized to display our trademark have direct access to us, which is further assurance that every attempt will be made to ensure that you and your children have the best information possible for dealing with losses of every kind.

Here are some of the ways that our symbol is used:

The Grief ❣ Recovery® Outreach Program
Certified Grief ❣ Recovery® Specialist
Grief ❣ Recovery® Certification Program
Grief ❣ Recovery® Personal Workshop
Grief ❣ Recovery® Program
The Actions of Grief ❣ Recovery®

Throughout this book, the phrase *grief recovery* appears many times. We have not added the heart ❣ each time. Instead we want you to use your imagination and put the ❣ symbol in with your own heart, to remind you what this book is all about.

With ❣,

JWJ, RPF, LLM

WHEN
CHILDREN
GRIEVE

PART ONE

Monkey See, Monkey Do

"My son's father died, and I want to know how to help him."

The above sentence may seem puzzling. It is an emotionally powerful statement of fact that raises many questions simultaneously.

And yet, "My son's father died, and I want to know how to help him" was the very real opening comment of a phone call Russell received at The Grief Recovery Institute. In order to understand the caller's specific circumstances, Russell had to ask the same questions that you might already be asking: Was her son's father her husband? Were they living together? Did she love the man? The caller answered yes to all three questions. As a matter of fact, this woman and her husband were very much in love, and she was devastated by his sudden death. In addition to the nine-year-old son, there were two daughters, one fourteen years old and the other five years old, about whom she was concerned.

Her husband had left the house one morning, an apparently healthy forty-year-old man. He arrived at work and suffered a massive heart attack. A chilling phone call informed her of his death.

Following his many years of experience, Russell encouraged the woman to talk about her relationship with her husband. But, with a singleness of purpose, she kept insisting that she wanted to talk only about helping her nine-year-old son. So Russell asked her to describe the problems she perceived her son to be having.

During the conversation that followed, Russell discovered that the boy was having many of the normal reactions associated with such a profound loss. But what was troubling the mom most was that her son would not talk at all about his reactions to his dad's death. She explained that when she asked him how he was feeling, he would say, "I'm fine!" and then clam up. When she asked a second time or pushed the topic, her son would retreat to his room and close the door. Russell said that he could imagine that might be pretty upsetting to her. After all, her son, whom she loved very much, had to be crushed by the death of his dad, yet he would not talk about it. This mom was sure that whatever was going on was not healthy for her son.

At that point, Russell recalled that earlier in the conversation she had said that she loved her husband very much. He said gently, "You told me that you loved your husband very much, and obviously you have been devastated by his death." After a short pause, in a small, choked voice, she answered, "Yes." Russell pressed on, asking her another question, one to which he was sure he already knew the answer. "When you and your son are together, and you get overwhelmed with emotions related to the death of your husband, what do you do?" Immediately she responded, "I have to be strong for him; that's what everybody tells me to do. So when I feel the tears coming, I go to my room."

A very long silence followed. Russell did not interrupt. Finally, and probably for the first time, she heard what she had just said. The lightbulb of awareness went off in her head. She understood. The silence ended with her saying, "Oh my gosh, he's doing what I do, isn't he?"

Later we will tell you more of the story about this mom and the nine-year-old boy whose father died. We will also tell you what happened with his two sisters. Their story will be an important illustration in helping you help your children deal with losses of all kinds.

WHY ARE YOU READING THIS BOOK?

You may be reading this book in response to a loss that has recently occurred.

You may be reading this book because you have observed something in your child that was caused by a loss that occurred some time ago.

This book may have been given to you by a loving relative or friend who has been helped by its contents or believes that it will be valuable to you in helping your child.

You may be reading this book in anticipation of a loss, one that appears to be inevitable.

The obvious reason you are reading this book is that you love your children and you want to be able to help them.

The important fact is that you have a genuine desire to do whatever it takes to help your child deal effectively with the experience of loss that has affected or will soon affect his or her life. We are honored to be your partners in ensuring that your child has the best possible information and the highest level of emotional safety as he or she deals with the wide range of emotions attached to losses of all kinds.

Establishing a foundation for dealing effectively with loss can be one of the greatest gifts you can give to your child.

CHAPTER 1

What's the Problem and Whose Problem Is It?

Because you are reading this book, there is a high probability that your child or a child in your care has experienced one or more losses. It is impossible to set down a list of losses that would have universal application to everyone reading this book. The following list represents the most common losses, in the sequence most likely to occur in a child's life.

Death of a pet

Death of a grandparent

Major move

Divorce of a child's parents

Death of a parent[s]

Death of a playmate, friend, or relative

Debilitating injury to the child or to someone important in
the child's life

The fact that one or more of the losses listed has occurred is only part of the problem. The other part is that you may not know exactly what to do to help your child deal with his or her feelings about this loss.

WHAT'S THE PROBLEM?

Something has occurred that is negatively affecting your child. You may be aware of this because of the ways in which your child is behaving. Many of the normal and natural signs of grief are fairly obvious. Most of those signs would be the same for a child's reaction to a death, a divorce, or some other type of loss. But for now, we will use a child's response to news about a death. Often the immediate response to learning of a death is a sense of numbness. That numbness lasts a different amount of time for each child. What usually lasts longer, and is even more universal, is a reduced ability to concentrate.

Other common reactions include major changes in eating and sleeping patterns. Those patterns can alternate from one extreme to the other. Also typical is a roller coaster of emotional highs and lows. As we mention these reactions, please notice that we are *not* labeling them as stages. They are simply some of the normal ways in which the body, the mind, and especially the emotions respond to the overwhelmingly painful information that something out of the ordinary has occurred. These reactions to a death are normal and typical even if there has been a long-term illness, which may have included substantial time and opportunity to "prepare" for that which will inevitably happen. We cannot prepare ourselves or our children, in advance, for the emotional reaction to a death.

This book (on behalf of your children) is about your child's reaction to death and other losses, and what you can do to help him or her. Because the topic of grief and potential recovery is so obscured by fear and misinformation, we are going to encourage you to examine the ideas you currently have about dealing with loss and to consider seriously whether those ideas are valuable for helping your child. We are going to presume that you are reading this book because you are eager to acquire the ideas and tools that will enable you to begin helping your child right away. So, let's get to work.

WHAT IS GRIEF, ANYWAY?

We have used the word *grief* several times in the opening pages of this book. Perhaps we should define the word for you, in the interest of clarity and mutual understanding. Many people associate the word *grief* only with physical death. We use a much broader definition that encompasses all loss experiences:

> *Grief is the conflicting feelings caused by a change or*
> *an end in a familiar pattern of behavior.*

As you'll recall, our list of losses included the death of a pet, death of a grandparent, moving, divorce of a child's parents, and death of a parent. Each of those losses represents a massive change or end from everything familiar. With death, the person or pet that has always been there is no longer there. With moving, the familiar place and surroundings are different. Divorce alters all of the routines in a child's life: it often includes changes in living situations and separation from extended family members and friends.

The losses we have listed carry with them the obvious emotional impact that we can all imagine would affect our children. But our definition of grief includes the idea that there are conflicting feelings. The concept of conflicting feelings requires a little bit of explanation. If you have ever had a loved one who struggled for a long time with a terminal illness, you may have had some feelings of relief when that person died. The relief usually stems from the idea that your loved one is no longer in pain. At the same time, your heart may have felt broken because he or she was no longer here. So the conflicting feelings are relief and sadness.

Moving also sets up conflicting feelings. We may miss some of the familiar things that we liked about the old house or neighborhood, and at the same time really like some of the things about the new place. Children are particularly affected by changes in locations, routines, and physical familiarity.

OBVIOUS AND HIDDEN LOSSES

Death, divorce, and even moving are obvious losses. Less apparent are losses having to do with health issues. A major change in the physical or mental health of a child or a parent can have dramatic impact on a child's life. And even though children are not usually directly involved with financial matters, they can be affected by major financial changes, positive or negative, within their family.

Society has identified more than forty life experiences that produce feelings of grief. At The Grief Recovery Institute we have expanded that list to include many of the loss experiences that are less concrete and thus are difficult to measure. Loss of trust, loss of safety, and loss of control are the most prominent of the intangible but life-altering experiences that affect children's lives. Intangible losses tend to be hidden and often do not surface until later in life, through therapy and other self-examinations.

As we move on in this book, we will explore in detail the most common losses that occur in the lives of young children. There is no predictable sequence to the occurrence of painful events. In fact, you may have been drawn to this book by an extraordinarily uncommon loss that has affected you and your children.

NEVER COMPARE LOSSES

"I cried because I had no shoes
until I met the man who had no feet."

This wonderful parable helps children develop a sense of proportion. It teaches them to look for things for which to be grateful. Unfortunately, it is often misconstrued to mean that when we have a loss, we must look for someone who has a larger loss, or more losses, so we won't feel so bad. Let us illustrate the misuse of that idea in its most heartbreaking form. Imagine a couple has had a young child die. They have two other children who are still living. This is what they often hear: "Don't feel bad, at least you have other children."

Well-meaning friends and relatives say such things in an attempt to help. But really what they have done is compared losses in order to minimize feelings. Do you think that having two other children diminishes the pain caused by the death of a child? The comparison, no matter how well intended, does the opposite; it makes the grievers feel worse. Worse, because the comment indicates that their friend does not understand what they are going through, which in turn leads to isolation, which further worsens the problem.

All loss is experienced at 100 percent. There is no such thing as half grief. This is particularly true for children. You have all seen a child howl when you take away a toy. The emotional response is immense, and the tears are real. As you begin to apply new ideas to the inevitable losses that occur in your children's lives, please remember never to compare losses and never to compare or ignore feelings.

TIME DOESN'T HEAL—ACTIONS DO

Shortly we will begin to address the six major myths that, if unchallenged, may limit your ability to help your child. Here is a preview of one such myth and its possible impact on your child.

We have been taught to believe that time heals all emotional wounds. The false belief that time heals is probably the single largest impediment to recovery from loss of any kind. Here is an amusing illustration: Imagine that you went out to your car and discovered it had a flat tire. Would you pull up a chair and sit down and wait for air to get back in your tire? Not likely. You would probably do one of two things—either replace the flat tire with the spare in the trunk, or call the auto club and ask them to come and change your tire.

In either case, the repair would be the result of action, not of time. Most people laugh when they hear the flat tire analogy. Some argue that there is a difference between recovery from the death of a loved one and the repair of a flat tire. We agree. But the only difference is that it takes different actions to change the tire than to complete the pain caused by the death or other loss.

> Recovery from grief or loss is achieved by a series of small and correct action choices made by the griever.

The fundamental purpose of this book is to teach you what those correct action choices are, and to show you how to use them to help your children. This book will provide very specific guidelines to help you accomplish the goal of helping your children. But before we get to the solutions, we must help you see what stands between you and the successful implementation of those actions.

NORMAL AND NATURAL

We hope the definition of grief we gave earlier will be helpful as you move through this book. Here is another statement that illustrates the fact that you already may be very well qualified to help your children:

> *Grief is the normal and natural reaction to loss. Of itself, grief is neither a pathological condition nor a personality disorder.*

That's a pretty powerful statement. It indicates that you don't need advanced degrees in order to help your child. And it would be wonderful if that statement alone could give us all the information we need in order to help. The problem is that while it is true that grief is the normal and natural reaction to loss, most of the information we have acquired during our lifetimes about dealing with grief is neither normal nor natural.

When you watch an infant respond to life, you see his or her natural reactions to loss. If a baby has something and you take that thing away, the baby may cry, loud and long. Many of you will remember the first time you left your child with a baby-sitter. Your child may have put up quite a fuss. Yet you knew that the baby's reaction to your leaving was very much within the range of normal behavior for an infant.

As a society, we seem to be willing to allow very young infants and small children the privilege of normal and natural reactions to loss, but as they grow older, we begin chastising them for being normal.

CRISIS BEHAVIOR

You may have heard the expression, "In a crisis, you return to old behavior." To illustrate the point from your own experience, think about any of the times that you've had an argument with your spouse or a close relative or friend. When the words and emotions start flying around, how often have you said or done something that you had promised yourself you would never say or do again? In the crisis of an argument or heated discussion, we often return to thoughts, feelings, and statements that we had hoped not to repeat.

Here's another example where your own experience can help you see how you may have carried ideas forward from childhood, and how they may have affected you and your children later. As a parent, have you ever opened your mouth and said something to your child, and then realized that the voice of your own mother or father had just popped out of your mouth? "You can't swim until an hour after lunch." There is a good chance that you heard that comment as a child. Yet have you ever stopped to ask yourself—or, better yet, a doctor—whether or not that warning is accurate? Quite often the ideas and language you use come from having heard them twenty, thirty, or forty years ago. You may not have consciously had that thought since you were a child, but in a crisis, you return to old behavior, or old beliefs. It is unlikely that, during the crisis itself, you will even question whether or not the behavior or belief is valid or helpful.

Please do not interpret what we have just said to mean that everything you heard your parents say is wrong. Most of what your parents taught you is helpful to assist you in living a safe and happy life. Instead, we are simply interested in examining those ideas that are not helpful for your children. We are merely suggesting that when certain situations arise, your brain automatically searches for information on that topic. Most of the information has been stored in your brain since childhood, waiting to be used in the appropriate set of circumstances. Unfortunately, most of the information you have stored about dealing with loss is probably not correct.

The mother in our opening story, faced with the crisis of her husband's death, relied on her stored memories to find an obsolete, incorrect idea: Be strong for your child. Her son, being a natural mimic, copied what he saw her doing.

Remember the questions posed in the title of this chapter, What's the problem, and whose problem is it?

The first part of the problem is that a child has experienced a loss.

The second part of the problem is that as the parents or guardians, in order to help your child, you may need better information than that which you yourself learned early on.

We know you are in a hurry to be able to help your child, so we are going to address the second part of the problem first.

BETWEEN THE PROBLEM AND THE SOLUTION: SIX MAJOR MYTHS

Before we can introduce the small and correct action choices that lead to recovery from loss experiences, we must discover exactly what is keeping you, and consequently your children, stuck. We are going to expose six myths and ask you to look at each of them and see whether or not they are helpful in dealing with loss. As you get a clearer picture of the ideas that do not work, you will be able to replace them with ideas that do work. These will become the new tools you will use to help your children.

Remember, the information children use to interpret their lives is passed down to them by their parents, teachers, clergy, and all others who have positions of influence in their young lives.

The things we teach are the things we ourselves learned when we were growing up. The first part of this book will ask you to look at some ideas that are almost universally accepted, but whose accuracy and helpfulness you may never have questioned. These myths are so familiar, there is a very high probability that you will recognize and relate to all of them.

The principle of learning by observing influential examples applies to all areas of life. Here we are primarily concerned with what children are taught to believe about dealing with loss. As you begin to recognize the ways that your children are learning from you, you may realize you learned a great deal of what you believe about dealing with loss when you were a child.

Before we begin to address specific losses like death, divorce, and moving, we are going to show you how the incorrect foundation for dealing with loss is established. The basic information upon which children rely to deal with loss is communicated very early in life. That information, right or wrong, tends to become the default setting that they will return to in response to all subsequent losses.

CHAPTER 2

Looking at Myth 1: Don't Feel Bad!

The logical response to the comment "Don't feel bad" should be "Why not?" Unfortunately, as a society we don't ask that question. Instead, we keep trying to reinforce the illogical idea that children shouldn't feel the way they feel.

Imagine that you have just hit your finger with a hammer, and you are jumping up and down, howling with pain. It is unlikely that it would be helpful for someone to come up to you and say, "Don't feel pain; after all, you didn't hit yourself on purpose." The comment certainly won't reduce the pain, the bleeding, or the swelling. Nor will being reminded that you didn't hit yourself on purpose make you feel any better.

Now, imagine that someone has just learned that his mother has died in an automobile accident. Among the very first comments he will hear from loving friends and relatives are: "Don't feel bad, she lived a long life." Or, "Don't feel bad, at least she didn't suffer." Or, "Don't feel bad, she's in a better place."

You have heard those kinds of comments. Perhaps you have even said some of them yourself. The fact is, you may never have been encouraged to look at them carefully. This means that you probably say the same sorts of things to your children. You want your children to be honest, yet you *unintentionally encourage dishonesty* in them by your incorrect reactions to their normal emotional responses to some of life's events. It is appropriate for them to have sad, painful, or negative reactions to sad, painful, or negative events. If you tell your children not to feel what they feel, you are inadvertently suggesting that they should be in conflict with the truth and at odds with their own nature.

Here is an example, from our previous book, *The Grief Recovery Handbook,* of how a child's normal response to a painful event is often reshaped by adults into what may become a lifelong, incorrect method for dealing with emotions.

A five-year-old girl has had her feelings hurt by the other children on the preschool playground. She goes home quite upset. She turns to Mom, Dad, or Grandma, and tearfully spills out her tale of woe. This healthy, normal expression and display of human emotion is met with the line that all of you already know is coming: "Don't feel bad. Here, have a cookie; you'll feel better." It doesn't have to be a cookie; even a healthy snack creates the illusion that we soothe feelings by eating.

The child has honestly presented an emotion, a sad feeling, to someone she trusts, a parent or guardian. The emotion is immediately dismissed—"Don't feel bad"—and then anesthetized with food. Think about it. Load a little body up with food and something will change. The truth is the child feels *different, not better.*

She is distracted by the cookie and the energy created by the food, but the painful emotions she experienced haven't been heard or talked about. Sometime later, the little girl wants to talk about the event at the playground, and she is told, "Don't cry over spilled milk." Again, the child and her emotions have been dismissed.

SWEET BUT DANGEROUS

The fact that the people who love us do not want us to feel bad is a sweet sentiment, but it's dangerous. A child is going to feel what he or she feels whether others approve of it or not. If the people around the child do not understand that the sad, painful, or negative feelings are normal and helpful, then the child will just go underground and hide his or her feelings. The child will begin to act fine, because that action is rewarded. "Isn't she brave?" or "Isn't he strong?" are the comments children hear when they cover up and bury their sad feelings after a loss.

WITHOUT SADNESS, JOY CANNOT EXIST

Feeling bad has a purpose. If you believe in the magnificent design of humans, then you must accept the fact that in order to have the capacity to feel happiness or joy, you must also be able to experience sadness or pain.

> Any attempt to bypass sad, painful, or negative emotions
> can and will have disastrous consequences.

One tragic by-product of the legacy of the simple phrase "don't feel bad," is that it often leads to a much worse cliché, "don't feel." Sadly, we have had to help too many people for whom "don't feel bad" turned into "don't feel at all."

WE ARE NOT EXAGGERATING

"Don't feel bad" (or "Don't feel sad) is the start of almost every phrase that suggests to children that what they are feeling is wrong. Here is a short list that most of you will recognize.

RELATING TO PET LOSS
Don't feel bad—on Saturday we'll get you a new dog.
Don't feel bad—it was only a dog [or cat, etc.].

RELATING TO DEATH

Don't feel bad—she's in a better place.

Don't feel bad—his suffering is over.

Don't feel bad—it's just God's will.

Don't feel bad—you did everything you could.

Don't feel bad—Grandpa's in heaven.

RELATING TO DIVORCE OR ROMANTIC BREAKUP

Don't feel bad—there are plenty of fish in the sea.

Don't feel bad—he or she wasn't right for you.

Don't feel bad—it was just puppy love [said to teenagers, especially].

RELATING TO CHILDREN OF DIVORCING PARENTS

Don't feel bad—this isn't your fault.

Don't feel bad—Mommy or Daddy will have more time for you.

Don't feel bad—Mommy and Daddy still love you.

Don't feel bad—you'll have two of every holiday and birthday [at Mommy's house and again at Daddy's house].

RELATING TO POOR PERFORMANCE AT SCHOOL

Don't feel bad—you'll do better next time.

Don't feel bad—you did your best.

Every one of the above examples begins with the phrase "Don't feel bad." This is not an exaggeration. It is a realistic representation of how it sounds in real life. A recent study determined that by the time a child is fifteen years old, he or she has already received more than twenty-three thousand reinforcements that indicate that it is not acceptable to show or communicate about sad feelings.

Just for fun, let's reverse this notion. Consider the phrase "Don't feel good," an idea you've probably rarely heard. The words look silly, don't they?

Imagine bringing home a report card with all A's and having your parents say:

Don't feel good—you'll do terrible next time.

Imagine telling a friend that everything is going very well in your
   life and having him say:

Don't feel good—things are going to get worse.

Imagine telling someone that you're in love and having
   her say:

Don't feel good—remember the divorce rate is 50 percent.

While there are some very negative people who are quick to remind us of life's pitfalls, the real issue here is that for the most part, we are allowed to have our positive, happy, and joyful emotions. Positive emotions are supported by the people around us. We rarely have to justify, explain, or defend our happiness.

On the other hand, sad, painful, or negative emotions are almost always met with Don't feel bad or You shouldn't feel that way. Remember our little girl who tells the truth about how she feels and the very first thing she is told is "Don't feel bad." Our question is, "Why not?" Why should we not feel bad when something bad happens to us? Why should we not feel sad when a sad event has occurred?

Why is it okay to feel good when something pleasant happens,
   but it is *not okay* to feel sad when something painful happens?

It is normal and natural to feel happy in response to positive events. It is equally normal and natural to feel sad in response to negative events. However, it is neither normal nor natural to dismiss any human emotions as not valid. The single largest source of emotional confusion in our society stems from the patently false idea that we somehow should not allow ourselves to experience sad, painful, or negative feelings.

When our children are infants, they communicate all feelings, happy and sad, at the top of their lungs. At a certain point, we start training children to believe that having sad feelings and communicating about those feelings is not okay. To the child's developing mind, it becomes a simple choice:

Happy feelings are good, and get rewarded
vs.
sad feelings are bad, and get punished.

From an early age, children learn the rules of the road. They are very clear as to which behaviors are rewarded and which behaviors are discouraged. That is why you need to make an objective evaluation about your own ability to understand or deal with feelings of pain or sorrow.

Just as you were watching your parents when you were young, your children are watching you now!

Earlier we mentioned how ridiculous it would seem if someone responded to a positive event in your life by saying, "Don't feel good." Well, even more absurd—not to mention destructive—is to lovingly tell a child "Don't feel bad" in response to a painful experience. As we suggested, to do so puts the child into conflict with his or her own nature, into conflict with the truth, and, finally, into conflict with the person who was attempting to comfort the child.

Ultimately, the same rules apply to both adults and children. If it is true that feeling bad has a valuable purpose, then this *truth* must be honored. Children, who naturally look to adults (especially their parents), for emotional guidance, are the ultimate victims of the misconception that avoiding or bypassing negative feelings can have positive results.

One of the unfortunate long-term results of the distorted ideas about dealing with sad feelings is evidenced by the number of people who show up at our seminars with the tragic complaint that they are unable to cry. Without exception, their stories contain a striking, repetitive message—"Don't feel bad."

Implicit in parenting is the idea of soothing our children. When an infant is too young to tell us in words specifically what is bothering him or her, we resort to general solutions, hoping that we can quickly make baby feel better. We put babies on our shoulder and rock them and coo to them. There is nothing wrong with this. Gradually we begin to recognize and differentiate our infant's calls of distress. We can tell if baby is hungry, or tired, or in need of a diaper change. We can then administer solutions to the child's specific problem.

And yet, in spite of our increasing familiarity with our child and his or her needs, there are often times when we cannot identify a specific problem, and therefore we cannot come up with a solution. But, as always, we do not want our children to feel bad. It is very difficult for us as parents to accept that there will be times when our children do not feel good, and when there is not a specific, identifiable, and correctable reason. It is very difficult for us just to allow them to feel bad, simply because that is the immediate reality for them.

Ask yourself the following questions: Are there times when you just feel a little down, a little blue, a little sad? If so, do you always know the exact reason? If you feel sad but don't know why, is that okay? As you think about those questions, imagine that you have just told a friend that you feel a little blue and that you have no specific reason. Then imagine your well-meaning and loving friend responding, "Don't feel bad, you should feel grateful—you have two arms and two legs and it's a beautiful day." Your friend doesn't understand that it's like comparing apples and oranges: your healthy arms and legs have no bearing on the sadness you are feeling.

We believe that a terribly misguided concept has evolved over the years; one that has led us to believe *that feeling bad is bad*. It is more helpful to understand that the essence of humanness is that our feelings are always changing, from one moment to the next. The most accurate illustration of the changeability of human emotions is seen in infants who go from happy to sad, or sad to happy, without any apparent external stimulus. They never question themselves until we start teaching them and showing them—"Don't feel bad."

WHO'S RESPONSIBLE FOR FEELINGS?

Many of you reading this book are aware of the victim mentality that seems to be almost epidemic in our society. If the word *victim* seems a bit harsh, substitute the word *helpless*. In either case, both concepts create the idea that no one ever seems to be responsible for what he or she has said, felt, or done.

Through our Grief Recovery Personal Workshops and Certification Trainings, we have observed how common it is during childhood for people to hear statements like: "Don't do that; you'll make your father angry," or "You make Mommy feel so proud." If you listen closely to those phrases, you will notice that they suggest that someone "makes" someone else feel something. A "victim" foundation is laid with the idea that *other people* are the chief architects of *our* feelings. Children are very, very smart. They realize quickly that if they have the power to make Daddy or Mommy feel something, then Daddy or Mommy can make them feel something.

Many parents, teachers, and other guardians will mistakenly say or imply that the words or actions of others are the primary cause of the child's feelings. The underlying message is that others made the child feel.

Here is a real-life story that happened to the child of one of our friends:

On a Monday, in a preschool class of four-year-olds, one of the two teachers was absent. The naturally curious children asked where the other teacher was. Here is the answer they were given by the other teacher. "You children were so bad last Friday, and did not listen, that Ms. X had to stay home and rest because of you."

What an incredible burden to place on the children! What an absurd illusion of power that the children could make someone else ill!

But wait, there are consequences in this story. The mom who told us this story was not telling tales out of school; she was actually at the school, in the classroom, and she heard the teacher say the exact words we quoted to you. The next morning, her four-year-old son, for the first time ever, said, "Mommy, I don't want to go to school today." After a little chat, Mom realized that he had been very much affected by what the teacher had said, and that he believed he had made the other teacher sick.

This incident highlights the long-term danger of passing on to children the idea that they are responsible for the feelings of others, which automatically sets them up to believe that others can be responsible for their feelings.

Teachers, like parents, are powerful influence sources in the minds and hearts of children. You or I may be able to dismiss ideas with which we do not agree. But children will hear the words of teachers or parents as gospel.

There is also a major issue of bad timing attached to the incorrect assignment of responsibility for feelings. Young children naturally live in the moment. They have a hard enough time understanding the concept of past and future. On Monday, the teacher asks them to remember their behavior last Friday. This is a recipe for disaster.

On Friday, both teachers missed the opportunity to help the children deal directly with their inattentiveness. On Monday, one teacher blamed the children for both teachers' inadequacies.

It would be wonderful if we could say that this was a rare and isolated incident. But we do not think that is true.

We want you to learn to communicate accurately about emotions, which your children can copy and learn. In simplest terms, comments like, "You really make me angry" need to be restructured. That statement makes the other person responsible for how you feel. It is more accurate to say, "I am angry." In this example, you are taking responsibility for your reaction to the other person's words or actions.

This can be one of the most powerful life lessons you teach your children. You must become aware of your use of the idea of others causing you to feel. As you change your language, your children will change, too. Then, everybody wins.

CHAPTER 3

Looking at Myth 2: Replace the Loss, Part One

When John was born, his family had a dog. Her name was Peggy, and she was a bulldog. She wasn't a puppy; she was about six years old. Peggy decided to adopt John when he came home from the hospital. Peggy insisted on sleeping in John's room, and rarely let young John out of her sight. As John grew up, she politely tolerated all of the annoying things that crawling children often do to dogs or cats. As John got older, they became best friends. They went everywhere together. They were a team. Peggy even taught John how to play fetch. John would throw a stick. Peggy, being older and a bulldog—not a retriever—would stand there, and John would run after the stick.

By the time John was six years old, Peggy slept in a basket in the kitchen. One morning John came into the kitchen and whistled to Peggy. Peggy did not budge. John whistled again. Still no response from Peggy. John ran to Peggy's basket, sensing that something was terribly wrong. In a moment, he realized that Peggy was not moving. He touched her. She was cold. John let out a bloodcurdling yell for the highest authority on Earth, "Mom!"

John's mother came running into the kitchen. John crouched over Peggy's basket with tears in his eyes. John's mom loved him very much, but she did not have any idea how to react to the emotional emergency caused by the death of John's dog. She said to

John, "The leaves turn to brown and fall to the ground, and the summer turns into fall." John looked up at his mom bewildered. He could not connect what had happened to Peggy with this science lesson. Seeing his confusion, she said, "Wait until your father gets home. He'll know how to help you."

Later that day John's father came home and learned what had happened. In this critical moment, at the death of John's best friend and companion, John's father said, "Don't feel bad, [pause] on Saturday we'll get you a new dog." Earlier we talked at length about the phrase "Don't feel bad." Here we go again, right on cue, a well-meaning, loving parent telling his son *not* to feel what he was feeling. And, promising that if he did a good job at not feeling what he felt, on Saturday they'd get him a new dog. We call this "replace the loss."

Nobody mentioned the feelings John was having over the death of his best friend. But, John's father, a primary role model in John's life, had told him, "Don't feel bad." John tried with all his might not to feel bad. But he *did* feel bad, so he tried to cover it up to show that he really was trying to do as his father suggested.

And true to their word, John's parents took him on Saturday to get a new dog. The problem, as John looks back on it, was that he was very much devastated by Peggy's death. He was not helped by his parents to deal effectively with the emotions attached to that loss. Without knowing why, John found himself unable to form an attachment with the new dog. He gave that dog to his younger brother.

But John was now saddled with two major misconceptions about dealing with loss: don't feel bad, and replace the loss. Tragically, this is a very common experience for children. The death of a pet and the subsequent ideas of how to deal with that loss create a model that becomes the unfortunate habit for dealing with future problems.

Imagine the long-term impact of these concepts—don't feel bad and replace the loss, when a parent introduces them to a child at a crucial, highly emotional moment in the child's life. The idea that we can mitigate the inevitable pain of loss by attempting to replace the lost object—in this case the dog—sets up tremendous conflict within the child. First, it dismisses the importance of the relationship between the child and the original animal. Second, it introduces the idea of disposability about valued relationships in general. Third, it creates an illusion that the child and the new dog will experience the same relationship that the child and his first dog had.

ALL RELATIONSHIPS ARE UNIQUE

Relationships with people, with animals, and even with prized possessions are unique. No two people can have identical relationships with anyone or anything. Yes, there can be parallels or similarities, but relationships are never identical—we bring our individuality into every relationship. Those of you who have more than one child know exactly what we mean. Each of your children is different and unique. Even parents who are blessed with identical twins know that as similar as they seem to be, the twins have distinct personalities.

Animals have different personalities, also. If you have ever been around a litter of puppies or kittens, you know that in very short order, you can tell which ones are bold, which are shy, which ones crave more attention, which ones are loners. When you combine the uniqueness of a child with the uniqueness of an animal, you recognize that the uniqueness of the relationship is irreplaceable. It could never be exactly the same with another animal because the animal would be different. So the notion of replacing the loss is a very dangerous idea, because it translates into replace the relationship, which is not possible.

What is possible is to have a new and unique relationship with a different animal. But, before we attempt to begin a new relationship we must complete the emotional relationship with the animal that has died. The same would hold true if the animal ran away or was given away. A failure to complete past relationships can make full participation in new relationships difficult or impossible. Children tend to try to re-create the old relationship with the new pet, mainly because they haven't been shown how to complete the prior relationship. It is unfair to the new pet, which is simply being itself, and yet is being pressured to be like the original pet. Remember, pets, like people, have different personalities, and trying to get the new dog to be like the old one puts it in conflict with its own unique and individual nature.

Sadly, we hear stories every day where replacing the loss becomes a wedge between people. It is common for loving, well-meaning friends, a day or two after the death of a pet, to show up with a new puppy or a new kitty. This gift is given without the prior knowledge, permission, or consent of the grieving pet owners. What you often end up with is an angry recipient and a bewildered presenter. All because our society keeps trying to convince us that if we don't feel bad, and if we replace the loss, we'll feel better.

Before we move on to another dangerous aspect of replacing the loss, let's make an observation about the joining of two myths. Don't feel bad and replace the loss almost always go together. Because we teach our children to believe that sad, painful, or negative feelings are not good, we automatically teach them to look for a different feeling, or relationship, or object to replace the sad feeling. We have observed many parents becoming frantic in an attempt to "fix" their children following the death of a cherished pet. What these parents failed to realize was that their children did not need to be fixed. Mostly they needed to be heard. Their feelings of loss were normal and natural; there was nothing that needed fixing, just affirming.

Children need to feel bad when their hearts are broken.
Don't try to fix them with a replacement.

THE STOLEN BICYCLE

It is not uncommon for one of the early losses in a child's life to be the loss of a possession. When John was about eight years old, someone stole his brand-new bicycle. He had all of the natural feelings that any child would have when something of his or hers is stolen. At home he was told, "Don't feel bad, [pause] on Saturday we'll get you a new bicycle." These remarks were accompanied by the lecture about being more responsible for his things. (How many of you remember that lecture?)

In any event, for John the ideas of how to deal with loss were now compounded. Once again it was: don't feel bad and replace the loss. John was a little boy; it did not dawn on him to question his parents' wisdom. Even though he did not feel good about the dog, and, later, the bike, he accepted his parents' advice as the truth. After all, where else would he learn how to deal with his feelings?

TOYS AND DOLLS—GONE BUT NOT FORGOTTEN

Many of you will have had the awkward experience of throwing away a toy belonging to your child, only to see him or her have a large emotional reaction. You may have been unable to retrieve the prized possession. In the escalating emotional climate, a sure way to make a bad situation worse is to tell your child that she shouldn't feel the way she feels. Remember our story about the little girl who was told not to feel bad after the other children had been mean to her? It is never helpful to tell a child not to feel the way she feels. To do so implies that there is something wrong with her for having a certain emotion.

Instead of directing your child away from his or her own emotional reality, it would be more helpful for you to share your observations about what might be more accurate for your child. "I can see that you are very upset that your doll is gone. You really loved that doll, didn't you? I'm sorry that I threw it away."

Chances are, you probably threw the doll away because it had been tucked in a closet or in the garage and your child had not seen or played with it for a long time. Yet the fact that it was out of sight or unused is irrelevant. Many parents make the mistake of insisting on an intellectual idea—i.e., "You hadn't played with the doll in months"—as opposed to acknowledging the child's emotional truth—"You really loved that doll."

The right and wrong of a story are intellectual concepts. The feelings associated with a story are emotional.

You must hear and acknowledge the emotions
before you address the facts of the story.

If a child brings the same issue up over and over, it is almost always because their feelings have not yet been heard.

IT'S TIME TO MEET LESLIE AND LEARN MORE ABOUT CHERISHED POSSESSIONS

You met Russell in the opening story as he helped the woman whose husband had died. You met John earlier as you read about his dog, Peggy, as well as other losses in his childhood. Now it's time to meet Leslie. She will tell you a story about her daughter, Rachel, and one of Rachel's cherished possessions.

"Rachel, like many children, had become very attached to her bottle. Between age one and two, Rachel's 'baba' was more than a source of food; it also represented a source of comfort and security. Of course, there were several babas. The babas went where Rachel went. As Rachel reached age two, it became time to prepare her to say good-bye to her babas. Being very aware of the loss-of-familiarity aspect of such a major change, I was not going to just make the babas disappear and let her struggle with her emotions. Plus, Rachel's younger brother, Justin, was still using a bottle, so there would be a constant reminder if I removed the bottle without any preparation.

"After introducing the idea that it was time to leave her babas behind, we created a plan. We talked about how many days it would be before it was time to say good-bye. We made a calendar for Rachel. At her age, Rachel loved stickers, so she used them to count off the days on the calendar. Then, I got a plain brown bag and Rachel colored it with her crayons and decorated it with more stickers and named it the Good-bye Babas Bag. We talked about her babas and how we could say good-bye to them.

"In the meantime, I had already begun by cutting out her afternoon bottle. Then I cut out her morning bottle, leaving her with only her evening bottle, which had the greatest significance for her.

"Rachel was very particular about wanting her babas cleaned and wanting the tops put on. She also wanted to make sure that they were *her* babas, and not to confuse them with the ones that were Justin's. We read books on what big sisters and brothers can do that little sisters and brothers can't do yet, which helped her feel big and important.

"I picked a day when I knew that our garbage would be picked up. Rachel proudly put the bottles into the special Good-bye Babas Bag and placed it into the garbage container. She watched from the window as the garbage collector took it away. Together we said bye-bye babas as the truck drove off.

"I was a little apprehensive, but Rachel was ready to go out and play. Later that night, she asked for her baba, and I reminded her that her babas had gone bye-bye. She started to cry. I asked her if her heart was feeling a little sad about missing her babas. She nodded yes, and curled up in my arms and held on tightly to the satin pillow she sleeps with every night. For about a week she would ask each night about her baba, and I would remind her that she'd said bye-bye to baba. After a few times she would remind herself about the bye-bye. And then she moved on.

"This event has been powerful in Rachel's life and has had benefits for her brother, not to mention her mom and dad. We all learned a lot. The emotional attachments children make are essential to happy lives. The effective completion of attachments is equally important. This became especially important in our family a few years later when we had to send Rachel's and Justin's dog to live with another family. The emotionally affirming lessons from saying bye-bye to the babas paid off again for the difficult task of saying good-bye to the dog."

REPLACE THE LOSS, PART TWO
(And a national divorce rate approaching 50 percent—uh-oh!)

With replace the loss as a backdrop, let's look at something that happens a little later in life, sometime during the teenage years. When our children become teenagers, nature encourages them to begin to experiment with courtship rituals, which leads to pair bonding, marriage, and mating. (Naturally we all hope our children will accomplish their life's goals at the right time—*and not too soon.*) But as parents of teenagers already know, it can be difficult to slow down the powerful forces of nature. As children begin to have romantic relationships and must deal with those feelings, they fall back on all of the information and misinformation with which they are already equipped. However, being young and not having as much practice, as do adults, with incorrect ideas, they will still try to do what is emotionally accurate—that is, they will try to tell the truth.

A teenager has his or her first romantic relationship. The birds sing, the sun shines, and the music sounds sweet. And then one day the relationship ends, and hearts are broken. Tommy comes home devastated. His body language indicates that all is not right in his world. Mom or Dad asks him what's wrong. He tells the truth, "Mary and I broke up." One simple, painful sentence. A perfect invitation for Mom or Dad to listen and hear.

But all too often, Tommy will be told: "Don't feel bad—there are plenty of fish in the sea." In other words, "Don't feel bad— replace the loss." Don't feel the way you feel—just go get another girlfriend.

The connection between the concepts "Don't feel bad—on Saturday we'll get you another dog" and "Don't feel bad—there are plenty of fish in the sea" is clear. The question is, do either of those comments help the child complete the painful feelings associated with the relationship that has ended? Or does the comment misguidedly encourage the child to bury the feelings about the first relationship and just go get another one?

What is unfinished about each past relationship gets buried but carried forward into the next one. By the time we marry, we may have a vast accumulation of unresolved grief about old relationships. It can become a veritable system of land mines, which is almost guaranteed to create explosions when a partner accidentally steps on a buried hurt. The first-year divorce rate is nearly 55 percent, and the long-term divorce rate has skyrocketed to almost 50 percent.

As we look critically at the two myths, don't feel bad and replace the loss, we can see the overwhelming damage they cause in our society.

Here's a thought:

> Maybe it's better to feel bad, when feeling bad
> is the normal reaction to an event!

CHAPTER 4

Looking at Myth 3: Grieve Alone

John enjoyed a close relationship with his grandfather who lived nearby. His grandpa taught him to hunt and fish and play ball. Since John's father was on the road a lot, Grandpa became very important to John. One day, while John was sitting in a high school class, a student came into the room and handed the teacher a note. The teacher read the note then walked down to John's desk and said, "John, your grandfather has died." As John remembers that moment, his mind began "Don't feel bad, on Sat—" . . . and he burst into tears, realizing that on Saturday he wasn't going to get a new grandpa.

The teacher, who adored John but was herself ill-equipped to deal with loss in the classroom, said, "John, maybe you should go to the office so you can be alone." So John went down to the office and sat by himself for what seemed like an eternity. Bear in mind that one of the most important people in John's life had died, and John was sent to sit alone as if he were being punished.

Finally, a family friend arrived to pick up John and take him home. He raced into the parlor to see his mother sitting in a corner crying. It was her father who had just died. John made a beeline for his mother, to both comfort and be comforted. He did not make it to his mother. One of his uncles stepped in and caught John and said, "John, don't disturb your mother, she'll be okay in a little while." The idea that she should grieve alone was cemented into

John's brain by the actions and reactions of the adults around him. John went to his room, too.

One of the dictionary definitions of the word *myth* is: an unfounded or false notion. By the age of fourteen, following the death of his dog, the theft of his bicycle, and then the death of his grandfather, John had acquired three very powerful myths or false notions for dealing with loss:

> Don't feel bad
> Replace the loss
> Grieve alone

If you identify with any of the three myths we have outlined, it is likely that you have been passing them along to your own children.

MULTIGENERATIONAL PASS-THROUGH

Multigenerational pass-through is fancy language for a very simple idea. What it means is that you teach what you know. Please remember who instilled those myths in John's mind. The input came from parents, teachers, and probably coaches and clergy, all of whom were powerful influences. And remember, too, that no one was looking at those ideas and asking if they were valid, true, or helpful.

Before we proceed, we want to help you understand the impact that the verbal and nonverbal actions of a parent or guardian can have on children. There is good and bad news concerning the ways children are affected by their parents. As stated earlier, the signals and information we pass along to our children are perceived to be absolutely true because, as little ones, they do not have any other information with which to make a comparison.

As children, we absorb a great deal of information during a time that is referred to as preconscious. Later in life, we are unlikely ever to know exactly why we believe some things and how we came to believe them. It is generally accepted that conscious memory begins somewhere between age two and age five.

A three-month-old baby does not understand nor can he ask why his mother is acting as if nothing has happened, or why she is "being strong" in reaction to a loss. Yet the three-month-old infant *will be* affected by the mother's feelings and actions. The memory experts tell us that the three-month-old will have no conscious memory of that time or of any specific events. However, at least two major things will have happened. One, the baby will have had some reaction to his sense of his mother's feelings. Two, the baby will store a memory about feelings, but may never have direct access to those feelings, nor what they mean or what caused them.

Children, from infancy on, are greatly affected by the actions and feelings of those who play important roles in their lives. As we get older, we often question many of the beliefs and values taught to us by our parents, our schools, our religions, and our overall society. In that kind of questioning, we begin to put a stamp of individuality on our belief systems, and that is a very good thing.

But most people never question the basic ideas and beliefs they have acquired about dealing with grief or loss. We often address a hall full of one thousand people and ask them to finish this sentence; "Laugh and the whole world laughs with you, cry and you cry . . ." Nine hundred ninety-five will sing out the word *alone,* proving our point that almost nobody questions those kinds of ideas.

In the story about the death of John's grandfather, everything that happened during the aftermath of that event suggested, influenced, and even commanded John to grieve alone. We apply this misguided idea to ourselves, we apply it to others, and, sadly, we apply it to our children.

Loss is inevitable. What doesn't have to be inevitable is the continuation of the problem of passing misinformation from one generation to the next generation, with no questions asked. The good news is that one of the major purposes of this book is to ask those

questions that can lead to better information with which to help
our children deal with loss.

GRIEVE ALONE—A CLOSER LOOK

Since this book is about the well-being of children, let's start with
infants to see if the idea of grieving alone seems true for them.
When infants are in discomfort of any kind, what is the very first
thing they do? They cry out for help. They do not "keep a stiff
upper lip." They communicate at the top of their lungs that they
are in need of some assistance. Loving parents respond to the call
for help and try to deliver it. Built into the parents' response is the
idea that the infant is indeed *not alone*.

Imagine then those same children, once they reach age five or
six, being told, "If you're going to cry, go to your room." In other
words, grieve alone. Or stronger comments, such as "Knock off
that crying, or I'll give you a reason to cry."

You can understand what this might do to children's sense of
trust: to be told, when they are having a normal and natural emotional reaction to a life event, that it is not okay to have that feeling.
Don't feel bad, and if you *insist* on having that feeling, by gosh, we
don't want to see it, so go to your room and grieve alone.

We are not suggesting that maturing children must return to the
state of squalling infanthood to get the comfort they need. What
we are saying is that it must be safe for children to express the
entire range of their emotions.

Naturally, as children grow, there must be a change in how they
communicate. It is the responsibility of the parent to understand
and guide that change. If you, as the parent or guardian, have
never questioned whether or not "don't feel bad" and "grieve
alone" are valuable concepts, then you might have some difficulty
in guiding your children effectively.

The fact that you are reading this book is a very strong indica-
tion that you want to help. You can't really help your kids until you

examine your own beliefs. If reading some of these ideas makes you aware that you have never taken a critical look at what you believe and are therefore passing on to your kids, please be gentle with yourself. It is not too late. As you change, your children will reap the benefits.

Since this chapter concerns the myth of grieving alone, we now suggest that you consider communicating about what you are learning here. This is an excellent chance for spouses to talk about ideas that they may never have discussed. If your spouse has died, perhaps this is one of the reasons you are reading this book—please be encouraged to talk with others, with family or friends. If you perpetuate the myth of grieving alone, both you and your children may suffer.

Recall the story that opened this book, that of the nine-year-old boy whose dad had died—he would get up from the table and go to his room and close the door—just as he had seen his mother do. If these were singular events, and there were no long-term conse-quences, we would not spend so much time debunking each of these dangerous myths. But these aren't singular events, and there are massive long-term consequences.

Do you think there might be an element of "grieving alone" impacting the divorce rate? If you guessed yes, we'd have to agree. How many of you, male or female, have not known how to deal with the emotions of a situation, and slipped off to your room, or to the garage, or to ride around in the car, or to clean up the kitchen, or to putter in the garden? Are those actions the direct result of a lifelong lesson to grieve alone? Do those actions lead to a strength-ening of the marriage, or to alienation and ultimately to divorce?

And how do those dynamics between spouses affect the children? Whether the children observe them directly, as with the little boy who saw his mom leave the table and go to her room, or indirectly, as the result of the emotional aftermath felt by each of the parents, the children are affected. Please don't delude yourself into thinking that you can hide things from your children. Even when they cannot identify what is going on, they know something isn't right.

WHY DO PEOPLE GRIEVE ALONE?

In simplest terms, people grieve alone because they are afraid of being judged or criticized for having the feelings they are having. Remember our first myth: Don't feel bad. This admonition suggests that we are somehow defective if we feel bad at all, or if the feeling continues for more than a moment. We are taught not to feel bad, but if and when we do, we must go to our room. We absorb the idea that it is not safe to feel bad, and, further, that it is not safe to feel bad in front of others.

If you think we are exaggerating, think about the following interaction between husband and wife. Husband comes home from work. Wife has had a very rough day, and nothing has gone right. Husband asks, "What's wrong honey?" Wife snarls, *"Nothing!"* Reverse the scenario, wife asks husband, husband snarls. This is not a gender issue. It is a safety issue. We are reasonably sure that almost everyone reading this book can relate to that little scenario. The question really is, Why does this woman, who loves her husband, not tell him the truth? The answer is that if she tells him how she feels, there is a very real possibility that he will say something like, [all together now] "Don't feel bad," followed by some intellectual cliché, like "Tomorrow is another day."

The fact is that she *does* feel bad. It has been a lousy day, and the one person she should be able to trust will immediately tell her *not* to feel how she feels. So, instead of telling him the truth, which would be unsafe, she grieves alone, and says nothing at all. The fact that she says nothing will probably create an additional gulf between the two of them.

Think again about the five-year-old girl in the beginning of chapter two who felt bad after having her feelings hurt on the preschool playground. Do you remember how unsafe it became for her to tell the truth about how she was feeling? Instead of being listened to, she is given a cookie. Could it be that little girl grows up to be the woman who says, *"Nothing,"* in response to her husband's question?

Why do people grieve alone? The long and short answer to the question is that people are taught from childhood onward that sad, painful, or negative feelings are not acceptable in public—or in private, for that matter. What you practice is what you perfect.

IS ALONE EVER OKAY?

In our speeches, lectures, and writings, we say grievers tend to isolate. We say that because it is true. It is true, but it is not natural. Remember, infants call out when they hurt. We don't isolate by nature; we isolate by training, by education, and by socialization. We isolate because we are taught that we laugh together but we cry alone.

Having said all that does not mean that we need to be surrounded by others twenty-four hours a day, seven days a week. Solitude is not bad. Collecting one's thoughts and feelings is not bad. Private time is not bad. There is a normal and natural need to keep our own counsel, away from others and a beehive of activity.

HERE'S SOME GOOD NEWS: DIFFERENT BELIEFS PRODUCE BETTER RESULTS FOR CHILDREN

So far we have taken a detailed look at three major myths and have discussed how they can have a lifelong negative impact on children and on their future as adults and parents. Now we want to take a look at the short-term and long-term benefits that have accrued to children who have been raised with better beliefs about dealing with loss.

Leslie's doctoral research involved comparing children who experienced the death of a family member. One group was made up of children whose parents or guardians had substantial awareness of the principles of grief recovery as discussed in our previous book, *The Grief Recovery Handbook*. The second group had no knowledge of the book or of any of its principles or recommended actions.

The essential difference between the two sets of parents/guardians was this: One group was helped to look at their own beliefs about dealing with loss (in much the same way that we have been doing so far in this book), while the other group was not. As the direct result of looking at and adjusting their own beliefs, these parents/guardians passed on better skills and ideas to the children in their care. The children in this first, relatively enlightened group, made much better transitions in the area of communicating about sad, painful, or negative feelings. They can talk safely and accurately about feeling sad, then move on to other feelings. The individuals in this group do not perceive themselves to be defective when they are sad. Over time, their bond of trust with their parents and their openness about their feelings has remained steadfast. This group includes children who ranged in age from four to eight when the major loss occurred in their lives. They have sustained the positive lessons learned as the result of their parents or guardians acquiring information about dealing more effectively with loss. Others in the study, who were eight or nine when their loss occurred and are now teenagers, feel incredibly safe in talking about or hearing others talk about loss. In addition, they are very helpful to friends who are dealing with loss.

Keep in mind that what you believe is what you teach. If you acquire more effective information about dealing with loss, you will be a better teacher. Your children will be the automatic beneficiaries. Personal stories illustrating those benefits appear in a later section entitled "Win-Win."

PAUSE TO REFLECT AND RECAP

As we proceed, you may be getting a clearer sense that what you learned as a child about dealing with the feelings associated with losses and disappointments was not correct and is not helpful. We want to caution you to be gentle with yourself, and to be gentle in your memory of those who taught you. You learned from individuals, from institutions, from film and literature, as well as from other sources. We know that everyone who taught you believed that what they were teaching was correct; otherwise they would have taught something else.

You may be growing aware that you are emotionally incomplete in your relationship with one or more of the people and events in your past. You may have started to realize that some of your own ideas are based on what was and wasn't shown to you in response to deaths, divorces, and other experiences in your life.

You may also have realized that you bought this book or were given this book so that you could better help your children deal with a specific painful event in their lives, but now you might begin to realize that the book keeps pointing to you first. If that is what you think, you are right. You might even be the teeniest little bit upset with us for aiming so much at you. We would love to say we are sorry, but we are not.

We are being as gentle as we can. We are not demanding that, before you help your children, you take full-scale action to complete all of the losses in your past. We are simply asking you to look at some crucial ideas and assumptions first, in order to help them. Someday, you may feel inclined to go back and deal with some of your past. When that time comes, you can get a copy of *The Grief Recovery Handbook* and get to work.

We have taken a detailed look at three of the major, life-limiting myths that are almost universally recognizable. We hope that you are beginning to have a new perspective about them, and will begin responding differently to your children. We have three more to highlight so that you will have an even greater awareness of what will be helpful for your children. You may be a little impatient for more actions, but hang in there, we are building a new and solid foundation. The next section may be a real eye-opener for you.

CHAPTER 5

Looking at Myth 4: Be Strong

We opened this book with the statement, "My son's father died, and I want to know how to help him," and the time has come to return to that story. The mom in our story was a loving mom. She wanted only the best for her child. But this mom had been socialized in a world that says you have to be strong. When she was around her son and experienced an emotion, she would immediately get ahold of herself so that she could be strong.

The myth about being strong is so big that it has two submyths, kind of like a planet with two moons. In the time since the death of her husband, well-meaning friends had bombarded her with another phrase, "You have to be strong for your children." Given the circumstance of her life, the fact that she was mostly in a daze, and the backdrop of a belief system that says you have to be strong, the idea that she might need to be strong for her children went unquestioned. It was a logical extension of the idea of being strong. Like most other parents, she had never stopped to question the wisdom or validity of that and many other clichés related to dealing with loss.

Another aspect of this situation is the simple idea that what you practice is what you perfect. Mom had practiced being strong for her son in other emotional events prior to the death of husband/ dad. In fact, she, like most of us, was so practiced that she was unaware that she was being strong. It took an outside observer to help her become aware that she was unconsciously teaching her son a habit that would hinder rather than help him in his life. Her son, in copying his mom's actions, was doing what he thought was the right thing.

WAIT, THERE'S MORE

As we observed, the boy imitated his mom who was being strong for him. And, if we shift our focus to one of her other children, we will see another variation on the be-strong theme. Mom's concern about her nine-year-old son was only the tip of the iceberg. Both of her daughters were also being affected by their mom's actions and nonactions, each in different ways.

Her fourteen-year-old daughter adopted the role of the family caretaker. Watching her mother's attempt at being strong, and her brother's silence, big sister decided that she was going to *save* everybody. In effect, she tried to transform herself, overnight, from a child into an adult. Part of this was copying her mom being strong for her son and daughters. Part of it was the accumulation of a young lifetime of misinformation from books, TV, and movies. And, being the eldest child, she had been told, "You have to be an adult now."

The daughter's attempt at "fixing" her family is also the flip side of the mom being strong for the children. The child learned "be strong for others." In all our years of working with grieving people, one of the most common and difficult-to-overcome problems is the child who was cast in or adopted the role of taking care of everyone else. It is one of the most heart-wrenching examples of loss-of-childhood experiences. While we are able to help people get their hearts back, we cannot give them their childhoods back.

We hope all the parents, guardians, teachers, and others who affect children will read this loud and clear: Please avoid phrases like "You have to be strong for your mom, or your dad" or "Now you're the little man, or the little woman, of the family." This is equally true whether dealing with the aftermath of a death or a divorce. We don't want our children to become little therapists. They still need to be kids. The death or divorce will affect them enormously without the additional burden of growing up before their time.

We have seen marriages sabotaged and destroyed by one partner taking care of the other in ways that rob dignity and integrity. This is often traceable to a child being strong for or taking care of parents or others, an unfortunate habit that turns into a lifelong impediment.

STRONG OR HUMAN, PICK ONE!

"I have to be strong for him; that's what everybody tells me to do. So when I feel the tears coming, I go to my room." Those are the words of the mom from chapter one of this book. The idea of being strong has become so distorted that it now implies that you should not have or demonstrate emotions in front of others, and especially not in front of children.

Perhaps it would be helpful if we redefined the word *strong* for you so that you can have a new perspective on what strong *really* is.

Real strength looks like this:
>   The natural demonstration of emotions.
>   Saying and doing what is emotionally accurate.

Real strength creates these results:
>   Teaches children how to communicate feelings,
>   not to bury them.
>   Sustains energy for other tasks.

As you can see, our definitions of strength really define *human*, not strong. It is very possible to be human and to accomplish what seems like an overwhelming number of tasks. The proper expression of emotions frees up energy to deal with life. The alternative is to hold on to feelings that, in turn, lead to explosions or implosions. Since our children are watching everything we do, we must become very aware that our beliefs and actions will become their beliefs and actions.

We have barely begun this book, and we have already identified several major misconceptions you may have learned when you were young:

> Don't feel bad.
>
> Replace the loss.
>
> Grieve alone.
>
> Be strong.
>
> > (Be strong for your children.)
> >
> > (Be strong for others.)

You may recognize yourself or your family in some or all of these myths. Again, we remind you not to be critical of yourself as we continue to uncover other beliefs that you may want to reconsider as they relate to your children.

CHAPTER 6

Looking at Myth 5: Keep Busy

Let's return to the mom from our first anecdote, this time as it relates to her five-year-old daughter. This brings us to another loss-related myth. Well-meaning family and friends had advised the mom to keep busy. Mom had become a beehive of activity. She scheduled herself into a frazzle, with the erroneous notion that the busier she was, the less she would feel the pain.

The keep-busy myth is so commonplace that we doubt there's a person who hasn't heard it countless times following all kinds of losses.

In this case, it was the five-year-old daughter who copied her mother's nonverbal communication and turned into a miniature whirling dervish. Although she had never been a particularly tidy little girl, she was now a junior Ms. Clean. She seemed to move at the same frenetic pace as her mom, as if somehow all that activity could help her not feel the pain in her heart. So, another major incorrect tool, "keep busy," had been passed on to the little girl.

Leaving aside the children for a moment, we can't begin to tell you how many widows and widowers have told us how exhausted they were as they tried to follow the advice of "keeping busy" that comes at them from all sides. Well-meaning friends and family, as well as clergy and therapists, counsel others to keep busy, having no idea how harmful such advice can be.

Grief, caused by death or by divorce, probably represents the largest change in the moment-to-moment life of a child. Adapting to life without someone who has always been there can be painful, difficult, and confusing. We grieve the change in everything that is familiar. Since the loss itself represents a massive change, we do not think it's a good idea for a child to make a lot of additional changes while still struggling with the upheaval caused by the loss. If the child was not a busy type of person before the loss, then becoming busy creates another huge change for the child to accommodate. And this is exactly what the child does not need: more changes. On the other hand, if the child was *always* a busy type, if busy is familiar, then we would not encourage any change in that style.

Imagine that this little five-year-old was her daddy's little princess. She had to have been crushed by his death. But, at five, she may have a very limited understanding of death. Her father's death created an unimaginable crisis for her, one for which she had no point of reference, no tools, no skills. She did have an incredible amount of emotional energy, the normal and natural reactions to her daddy's death. All she could do was watch her mom, and copy her mom; consequently, all that emotional energy was diverted into cleaning and keeping busy.

The good news for the little girl, and her sister and brother, was the fact that Mom called The Grief Recovery Institute very shortly after her husband's death. As Mom got help from us, she was able to guide her children. We are pleased to tell you that Mom and the three children are doing well, even though all of their lives were forever altered by the death.

A DANGEROUS ILLUSION

Keeping busy, in addition to being exhausting, can create a dangerous illusion. The illusion is that, as you throw yourself into activity, and days, weeks, and even months pass, you have actually done something constructive to deal with the unfinished emotions that are naturally attached to death, divorce, and other losses.

Nothing could be further from the truth. All you have accomplished is to distract yourself from the pain caused by the loss, and, in the process, there is a high probability that you have buried the emotions further out of sight.

But the emotions of grief are powerful. They do not fade away so easily. You have probably known people who will talk about events from thirty, forty, or fifty years ago with an emotional pain that makes it sound as if it happened yesterday. You can stay busy all day, but when you finally stop and lie down at night, you're liable to find the same pain in your heart that was there yesterday and the day before.

In this regard, children and adults are identical. Keeping busy doesn't have any more positive value for them than it does for adults. Children may be a little more honest, more willing to tell you they are still hurting. Please listen to them. Don't encourage them to steamroller over their feelings just because you may have adopted the myth about keeping busy a long time ago.

THE REAL IMPACT OF LOSS: KEEPING BUSY AND DWELLING ON PAIN

Please keep in mind that the death of a loved one is a staggering event in the lives of everyone concerned. It is not uncommon for those affected to have a sense of being numbed or even paralyzed by the event. It is often as if a machine has overloaded and a circuit breaker, sensing the problem, has shut off the equipment. As people respond to the overwhelming emotions caused by loss, it would be foolish to rush them back into the mainstream of life while they are still trying to comprehend this new reality.

So much of the well-intentioned but misguided help comes from the faulty interpretation of clichés and stereotypes. "An idle mind is the devil's workshop" is the basis for the idea that it's best to keep busy. It is easy to see how that adage could be applied to a loss situation in an extremely damaging way.

We know that the human body can produce powerful chemicals to act as an anesthetic when its owner has experienced a physical calamity. Our bodies will even force us to lose consciousness in order to protect us from unmanageable pain.

There appears to be a psychological equivalent that shuts us down upon the receipt of news that is too painful to accept. This is especially apparent when people are notified of the sudden death of a loved one.

Most grieving people report a sense of numbness, either constant or intermittent, for a period of time following the death of a loved one. The length of time is unique and unpredictable for each person. With all that we have observed at The Grief Recovery Institute, there is no doubt in our minds that this numbness has a positive function. Most obvious is the way this mechanism lets our brains, our hearts, and our spirits accommodate and accept the unwanted reality of the death. And the numbness can literally take us out of the busyness of daily events so that we can deal with the grief directly and not distract ourselves with activity.

If all of the above is true for adults, it is exactly the same for children. In some ways coping with sudden tragedy may be worse for them, because they may not have the communication skills to explain what they are thinking or feeling. If the adult would respond to this event by being preoccupied and unproductive at work, so the child would respond by being preoccupied and unproductive at school. The problem is that schools tend to treat grief as a discipline problem rather than as an emotional reality.

We want to reemphasize that the degree to which each person—adult or child—is impacted by loss is unique and individual. We are talking about the time in which people are not very functional, where work or school would not make much sense to them. There are many factors that dictate an individual's reentry into the mainstream of life following loss. There is no realistic, universal time line that we would feel safe committing to print.

The three important areas that are affected by grief are emotional, spiritual, and intellectual. Grief, by definition, is the emotional response to loss of any kind. We alert you to pay particular attention to that aspect of your child's life. As the parent or guardian, watch all three elements in your children—emotions first, spirituality second, and intellect last.

Let us take a moment and discuss the word *spiritual* in the context of grief. While we are not implying any specific religious connotation when we use the word *spiritual*, we are aware that many people use religious principles as part of their concept of spirituality. When we talk about spirituality, we are referring to the soul, or the spirit, or the intuitive part of the human existence. We are talking about that aspect that cannot be defined as emotional or intellectual, but that requires a category of its own.

We also believe that you, as the parent or guardian, have the clearest understanding of the spiritual aspect of your children. We strongly suggest that you stay alert to any differences or changes you may perceive in that realm of your child following losses of any kind. In addition to being attentive to their emotional ups and downs, try to stay aware of their spiritual condition. We know that at the same time you may be preoccupied with your own emotions and your own spirit. It may be your mother, father, brother, sister, or spouse who has died. It will be difficult enough for you to follow your own roller coaster of emotions, much less that of your child. Even though you may feel split between your own reactions and your desire to help your child, it is important that you try to use the experience to create a connecting bridge to your child rather than letting it become a gulf.

DWELLING ON PAIN IS SOMETIMES THE RESULT OF NOT BEING HEARD

There may come a point at which you realize that one of your children has gotten stuck, which may be caused by a variety of reasons. One of the reasons could relate to a comment we made very early in this book. Again:

The right and wrong of a story are intellectual concepts. The feelings associated with a story are emotional.

> You must hear and acknowledge the emotions
> before you address the facts of the story.

If a child brings up the same issue over and over, it is almost always because his or her feelings *have not yet been heard*.

It is the very last phrase of that sentence that is most important here. When a child appears to be dwelling on an issue, it may very well be that he or she does not feel heard. Children repeat what they are thinking and feeling, with the hope that someone will acknowledge the truth of their communication. But if the child's parents or guardians are saying things like, "Don't feel bad," or "You have to be strong," or "Just keep busy," the child will certainly not believe the adults are listening or hearing.

Some children will keep hammering away, desperately trying to be heard. You might mistakenly conclude that they are dwelling negatively on the topic. We, on the other hand, would think they are very smart and are unwilling to give up. Many children, after a few attempts at being heard, will quit trying. They will bury their feelings, and they may develop some behavioral problems to accommodate the energy that is generated and regenerated by the lack of acknowledgment of the emotions they are trying to communicate. Sadly, when this happens, the original grief incident is often overlooked. The focus is shifted to the child's behavior, which is seen incorrectly as a discipline problem rather than a grief issue.

HEARD AT LAST

Occasionally we are called upon to consult with movie or television producers and directors on the subject of grief. They want to know how a character might react given the circumstances that surround a loss. We are always happy to get these requests because it means that some aspects of grief and recovery will be portrayed correctly. Still, our input makes only a small contribution to counteracting the enormous amount of misinformation transmitted in fictional dramas, as well as on news broadcasts.

John remembers one of these experiences vividly, and we will use his experience to demonstrate a point. John had been invited to consult with a well-known producer and a well-known director on a film project they were doing. They met at a restaurant high in the hills above Los Angeles. John arrived first, the director next, and finally the producer.

The producer was accompanied by his nine-year-old grand-daughter, Briana. When they arrived at the table, the producer explained, somewhat regretfully, that he was baby-sitting that day and he apologized for the young girl's presence at the meeting. John remembers thinking, I'm glad she's here; at least there will be someone fun to talk with. The consulting portion of the meeting took only about a half-hour. Then the director and producer moved to a nearby table to discuss what they had just learned.

John and Briana began to chat. In a very short period of time, Briana decided that it was safe to talk about feelings with John. Seemingly out of the blue, and in a very serious manner, she stated, "Mr. James, you know I've had a very hard life." Rather than smile condescendingly, John asked, "What has happened?" Briana began to tell her story. "My other grandpa died. We had to move from New Jersey to California; and I had to leave all my friends, and I don't want to make new ones. And my cat, that I've had all my life, ran away."

Remember Briana's opening statement? "Mr. James, you know I've had a very hard life."

Had Briana had a very hard life? Most would answer that, yes, she's had a pretty hard time. In truth, though, her story is fairly typical. For a nine-year-old to have a grandparent die, to experience a move that entails leaving friends and starting in a new school, and to lose a pet, is not unusual. So if the circumstances of her life were not unusual, what made it seem to her that she has had "a very hard life"? When you think about it in relation to what you have already read in this book, the answer may become clear. What made it seem like such a hard life to our little girl was that she could not seem to find anyone who would listen to her. No one could hear her cries for help. No one had heard and responded helpfully to her feelings about the events.

During the conversation John learned that this failure to be heard was not from lack of trying on Briana's part. She had tried to talk about her feelings concerning the changes in her life quite a few times during the first couple of months following all of the upheaval. But every response she got in reply was intellectual in nature, or was preceded by the typical comment "Don't feel bad." Everyone, it seems, had tried to fix her head by explaining the facts of the situation. But she understood the facts. Her head wasn't broken—her heart was.

When she heard John talking to her grandfather and the other man about dealing with loss, she realized that John might be the "listener" she needed. So Briana's comment, "I've had a very hard life," wasn't out of the blue. It was the result of observing that John talked openly about sad things.

CHAPTER 7

Looking at Myth 6: Time Heals All Wounds

This might be the single most dramatically inaccurate piece of misinformation that has been imposed on all of us. Like most false beliefs, this idea has a partial basis in reality. Recovery from loss and completion of emotional pain do happen within a framework of time. However, there is a world of difference between time healing a wound and a wound healing within time.

We have presented our humorous explanation of the fact that time, of itself, does not and cannot heal. We asked you to imagine that you'd gone out to your car only to discover it had a flat tire. We wondered if you would pull up a chair, sit down, and wait for air to get back in your tire. The answer was obvious. Time was not going to put air into your tire.

Why, then, does the myth that time heals persist? Let us try to help you see how an aspect of truth supports the falsehood. Major loss events like death and divorce can produce an overwhelming amount of emotional energy. As we noted in the last chapter, sometimes the pain is so unbearable that our hearts and brains become numb. As we accommodate and accept the reality of the loss, some of the pain will diminish naturally. Most people interpret that reduced pain to have been caused by the passage of time. That is accurate, but only in regard to the immediate pain associated with the loss.

In the last chapter we talked about the dangers of the myth that advises us to keep busy. If we combine *keep busy* with *time heals all wounds,* the resulting idea is that if you just stay busy enough, more time will pass, and eventually you will heal. Many of the myths are hooked together with almost invisible wires, as in the illustration above. Since most people never look closely at these kinds of ideas, they usually don't realize what those comments mean. For example, while there are people who would never advise grieving friends that time heals all wounds, they would, without thinking it through, counsel them to keep busy.

CORPORATE GRIEF AND GRIEF IN THE CLASSROOM

If you fall down and suffer a severely broken arm, you might get four to six weeks off work with disability pay, and everybody would joyfully sign your cast. What happens when your mom dies, or your husband, or brother, or child? What is the nationally accepted average time off to deal with the overwhelming feelings caused by the death of a loved one? The answer: three days! Yes, three days, for a broken heart caused by the death of a loved one, and several weeks off for a broken arm. Is it possible that our priorities are a little bit out of whack?

If societally accepted ideas imply that you should be back at your desk four days after the death of a loved one, looking good and being productive, then the myth that time is an element in recovery is reinforced. It makes very little sense that someone might be able so quickly to recover his or her equilibrium following the death of a loved one. Yet, the idea of "instant recovery" adds a burden so powerful that many grieving people are inclined to say "I'm fine," even though they are drowning in a sea of sorrow.

While we have used the adult-oriented example of time off from a job, we ask you to consider this notion as it applies to your children. Are the same kinds of ideas used in the workplace also used in the school setting? Are our children expected to be back in class,

bright-eyed, cheery, and productive only a few days after a shatter-
ing loss? Sadly, this is often true.

We know that teachers are required to have skill and awareness
in many academic and administrative areas, each of which can be
important to the well-being of children. But we also know that
teachers are bound to interact with grieving children on a regular
basis. Therefore, we believe that schoolteachers, counselors, and
administrators can benefit greatly from the ideas in this book.

NO TIME ZONES

Many well-meaning organizations publish guidelines that suggest
time frames for individual losses. They suggest that it takes one
year to get over the death of a relative or friend, two years to get
over the death of a parent or spouse, and they believe that you
never get over the death of a child. Let's look closely at the phrase
"get over." Getting over implies forgetting someone. There is no
way that you will ever forget your mother, or your spouse, or your
child. We have known people who have been told, when they dis-
played some sad emotions about someone who had died some time
ago, that it was proof they hadn't gotten over them yet. By that
logic, if you have a fond memory of someone who died a long time
ago, you are equally not over them.

We do not agree that time, in and of itself, will help you. We
have met people who have been waiting for twenty or thirty years
or more to feel better. One of the common expressions that grieving people hear is, "You should be over it by now." When people
don't feel better after some arbitrarily allotted time period, they
begin to say to themselves, "I should be over it by now." Unfortunately, the myth that time heals keeps reinforcing the mistaken
idea that time is an active force with the power to do something,
and if you just wait long enough, you'll be fixed.

Finally, the idea that people *never* get over the death of a child is perhaps the most damaging piece of misinformation we have ever heard. We have watched people literally stop living as the result of having been told that they won't get over the death of their child. Why bother going on if you are never going to feel better?

The good news is that when we have the opportunity to communicate directly with people, even though they have been brutalized by the statement "You never get over the death of a child," we can give them hope by helping them use different language. We tell them that they will never forget their child, and if they take the actions of grief recovery, they will be able to retain fond memories without those memories turning painful. They will be able to regain a life of meaning and value, even though their lives are vastly different than they might have been had the death not occurred. It is important to equate this same set of ideas in relation to the fact that your children may also be hearing that language, "You never get over . . ." They will automatically insert "death of brother or sister" into what they are hearing. Children need the same corrected language as adults so they don't get caught in the trap of believing that they will never feel different or better.

We are not saying that the completion that results from taking the actions of grief recovery means that you would never miss someone who is no longer alive, or that you will never be sad again. Instead, as you begin to recover, an entire range of emotions will be available to you when you are reminded of someone who is no longer here. Keep in mind that you are sometimes sad about and miss someone who is still alive but who lives across the country. What must be altered, on behalf of your children, is the idea that sad is bad. Sad is not bad; sad is just part of being normal.

Many organizations tend to rank losses into a hierarchy based on the specific label of the relationship. This troubles us. Since all relationships are unique, we think it's dangerous to rank them. Those of you who have several children will know that, while you love each of them, each relationship is distinctive and individual. We must be especially careful to remember that we are not grieving our relationship with a title (that is, mother, spouse, or child), but instead we are grieving our very special and individual relationship with that person who has died.

We have spent a lot of time on this topic for a couple of reasons. First, we want you to have better information for your children. Second, there is a possibility that you have had a child die, and that is one of the reasons you have this book. You may be very concerned about your other children and how they might be affected by the death of their sibling. We want to alert you not to get caught in the trap of comparing or ranking losses.

PART TWO

Moving from Grief to Recovery

So far, we have focused on the fact that misinformation is unconsciously passed from generation to generation. We hope you will now feel encouraged to acquire more accurate information to pass on to your children.

We explained that grief is the normal and natural reaction to loss and defined grief as the conflicting feelings caused by the change or end in a familiar pattern of behavior. The two statements are wonderful and simple ideas about what grief is: the appropriate reaction to loss of any kind.

Grief is important. But it is not really the topic of this book. This book is about recovery from and completion of the pain caused by the grieving experience. Specifically, it is about helping you help your children deal with the particular loss that attracted you to this book. It is also about helping you furnish them with effective ways of dealing with any of the losses they will encounter throughout their lives. The raw emotions caused by loss stand alone and are specific to every individual. You have probably heard that everyone grieves in his or her own way and at his or her own pace. That is true. It is not our place to tell anyone how to grieve. Rather, it is our task to help people deal with the aftermath of loss.

CHAPTER 8

Looking for "The Book"

In 1977, John was a contractor building solar homes. He and his wife had a two-year-old daughter and were eagerly awaiting the arrival of another child. But all did not go well, and the baby boy only lived three days. John was devastated.

While John had experienced many other losses, he was totally unprepared for the pain caused by the death of his son. He went to every imaginable source for help, and at every one he was met with comments like "Time will heal," or "God won't give you more than you can handle," or "Just keep busy." But his grief was unbearable, to the degree that he very seriously considered taking his own life. He was thirty-four years old. He realized that by the law of averages he had at least another thirty or forty years on this planet. He knew that he would not be able to live with that kind of overwhelming pain for very much longer. He made a vow to find a way to feel differently.

At first, John haunted the bookstores. He was at the doorstep when their new shipments arrived. He was looking for a very particular book. He did not know the title of the book, but he knew he would recognize it when he saw it. He was sure, with the millions of deaths that occur every year, someone would have figured out what to do about the constant pain caused by death of a loved one. He never found the book.

He did find hundreds of books that told him, sometimes in very flowery terms, how he felt. The problem was that John already knew how he felt. He needed no assistance in recognizing the inordinate pain that was his constant companion. One virtue of those books was the soothing idea that he was not alone in feeling bad. While this information was valuable, it was also limited; it still did not explain what he could do to begin to feel differently.

When the book search failed, John set off on an odyssey to discover what, if anything, he could do to help himself deal with the pain and recapture a willingness and desire to participate fully in life. He followed every idea he heard of or could think of, until he had patched together several simple but helpful actions. First was the need to examine all the other losses in his life—death of dog, death of grandpa, romantic breakups, Vietnam, Dad's death, younger brother's death, divorce—and uncover all the ideas he had ever learned about dealing with loss.

In part one of this book, we looked at the six myths about dealing with loss. John's story reveals how the information he gleaned during childhood became the basis for the ideas he would rely upon for the rest of his life. It was not until he reevaluated his assumptions about dealing with loss that he could begin to do anything differently.

John realized that he had to look carefully at and finally dismiss these very powerful ideas: don't feel bad, replace the loss, grieve alone, be strong, keep busy, and time will heal. That was difficult enough since those ideas are so ingrained in our culture. To compound the problem, there was no clear alternative or support for doing anything other than clinging to those myths. What was even more difficult was the question, If you take away those ideas, what do you have left? It can be more than a little scary to give up something and not have something with which to replace it. Still, somewhere between the old myths and some of the new ideas he was discovering, John asked himself the question that led to the solution.

John had already experienced some major death losses in his life. The losses were very painful for him, but he had somehow managed to rebound, although he could not have told you how. He wondered, "Why is this loss so different from the others?" and "Why am I not able to rebound like I did after my younger brother died?"

The key word in the first question was the word *different*. As John thought about what had been different, he also came up with a couple of other words—*better* and *more*. The question began to evolve. Now he asked himself, Were there things that he wished had been different, better, or more in his relationship to his son? The answer was yes. Obviously, he wished that the medical circumstance of his son's life had been different, and within that were nagging questions such as, "Should there have been more prenatal visits to the doctor?"

The idea of different, better, or more began to help John uncover some of the sources of emotional energy that were contributing to his pain. The fact that we all will look back over relationships that have ended or changed, and discover things we wish had been different, better, or more, appears to be an almost universal truth. In the rare instances when someone insists that there is nothing that could have been different or better, they most assuredly wish there had been more.

John started examining his relationship with his son who had lived for only three days. He had spent the better part of nine months visualizing events and anticipating emotions that he hoped would happen in the future. Like any other parent-to-be, he had plans for what he would do if the baby were a boy, and plans for if the baby were a girl, and no matter what, he was going to make sure that this baby was the most loved baby of all time. Every thought and feeling he had was normal and natural. This is what people do—generate hopes, dreams, and expectations for the future.

As John's odyssey toward recovery continued, he began to realize that the pain in his heart had a lot to do with the fact that the baby's death had ended all of John's hopes, dreams, and expectations for the future. It is obvious that a beginning relationship with a child would automatically include many thoughts about the future.

But the idea of broken hopes, dreams, and expectations is not at all limited to the death of a child. Hopes, dreams, and expectations pertain to all relationships, even to those that were not always happy and fulfilling. When someone with whom you have had a stormy and checkered relationship dies, it ends the hope that you will ever be able to repair the relationship. Your children will experience losses in which much of the unresolved grief will relate to the fact that the relationship ended with many unrealized hopes, dreams, and expectations.

This is particularly true of divorce. Divorce by its very occurrence implies broken hopes, dreams, and expectations about the future. Children are devastated by the consequences of divorce that concern their futures. Children sometimes get caught up in ideas that relate to a false notion of things that they might have done differently, better, or more, to change the outcome of their parents' marriage. It is important for you to help your children see that they were not the cause of the divorce. Just telling children that they were not the cause is not enough. Having more helpful ideas about different, better, or more and about broken hopes, dreams, and expectations will assist you in helping your children deal with the enormous emotions created by the divorce of their parents.

You may have noticed that we have used the phrases "different, better, or more," and "broken hopes, dreams, and expectations" several times over the past few pages. This is intentional. We want you to start incorporating that language into your thinking. As part two unfolds, we are going to explain how you can use the ideas contained in those phrases to help your children deal with the loss that brought you to this book.

JOHN'S QUEST CONTINUES

As the direct result of delving into his prior losses, John was freed to look at the relationship with his child who had died from the perspective of different, better, or more, and to consider his own unrealized hopes, dreams, and expectations. The discoveries that

John made were very emotional. At first, he did not know how to handle the thoughts and feelings that had emerged from his search. The next part of his journey was to learn how to convert his discoveries into a practical format that would allow them to be communicated and completed.

John's personal success at discovering and completing the emotions that had been incomplete and uncommunicated helped him recapture a sense of purpose in his life. Although his life had been affected and changed by his son's death, he felt prepared to go on, finally feeling better, reenergized, and ready to get back to the business of building solar homes.

But fate seemed to have a different plan for John. Friends, hearing of his ability to rebound into life, started introducing him to other parents who had children who were terminally ill or who had just died. After a while, John was spending more time helping grieving people than building houses. In rather short order, well-meaning friends figured out that if John could help people who had had a child die, maybe he could help people who had had a spouse or a parent die.

Soon it became obvious to John that he was no longer in the construction business; he was now in the emotional business of helping other people deal with the pain caused by every imaginable type of loss. And the more people he helped, the more people showed up who wanted and needed help. And it got so large that he could not help all of the people who needed help; he realized that as just one man, there was only so much he could do in any given day.

So he sat down at what would now be considered an archaic computer/word processor, and he wrote the book that he had never been able to find. He had not been able to find it because it had not been written yet. The original version of *The Grief Recovery Handbook* was written on a primitive computer, self-published, and distributed to anyone who was looking for the answer to the same question John had asked years earlier: I know how I feel; what do I do about it?

Shortly after John self-published that first version, he and a colleague, Frank Cherry, made a deal with HarperCollins, and in 1988 they published a jointly authored version of *The Grief Recovery Handbook: A Step-by-Step Program for Moving Beyond Loss.* In 1998, HarperCollins published a revised edition by John and Russell, *The Grief Recovery Handbook: The Action Program for Moving Beyond Death, Divorce, and Other Losses.*

John's frustration at not finding a book that could explain what actions he might take to complete the pain caused by the death of his son has led to the repair of millions of hearts damaged by losses of every kind. And it has led to this book.

The actions that helped John complete what was emotionally unfinished for him, about his son who had died, are the very same actions that you can use to help your children deal with the emotional pain caused by the losses in their lives.

CHAPTER 9

What Is Incomplete Grief?

We keep mentioning that the task is to grieve and complete the relationship with all that is emotionally unfinished at the time of a loss. What exactly does that mean? Grieving occurs automatically, evidenced by a lowered state of feeling, confusion, and the possibility of every human emotion, sometimes many at the same time. Completion, however, is the result of specific actions.

Let's take a look at a real-life situation. Two women, Nancy and Mary, are good friends. They spend a day together. Near the end of the day, they have an argument. They exchange harsh words and part company. When Nancy gets home she doesn't feel right; she is unsettled by the incident. We would say that she feels incomplete. What does she need to do for herself so that she can feel complete? There is an obvious action that would create completion, and it is this: Nancy picks up the phone and calls Mary. She says, "Hi, Mary, it's Nancy. I'm sorry, I got snappy with you. I apologize." Mary responds, "Thanks, me, too. I'm sorry I was a little edgy today."

In that simple example, both friends become complete by taking the action of apologizing. It just so happens that Nancy went first, by making the call. In that scenario, the two are obviously good friends who had just had a little flare-up. The apologies are small and correct action choices that lead to completion of the event that caused the separation. They are now free to continue their friendship and do not have to carry the unfinished emotional business of that event into their next interaction.

Imagine the same story with a different ending. The original squabble happens, but neither woman picks up the phone to apologize. Two days later one of them is killed in a traffic accident. It is obvious who will be left with undelivered emotional communications about the last interaction in that relationship. That woman is left emotionally incomplete, because of the unexpressed communication that would have resolved the spat. You could say that she is "emotionally incomplete" about the incident, or that the undelivered communication represents one element of "unresolved grief."

We have talked at length about the phrases "different, better, or more," and "broken hopes, dreams, and expectations." The very first thing that leaps out at you is that the surviving woman certainly wishes that the last communication with her friend had been different. Imagine her standing at her friend's funeral, her mind replaying the argument over and over as she asks herself thousands of times why she had been so stubborn and hadn't called to apologize.

We are not suggesting that if she had made the call, as in the first scenario, she would be feeling any less sad at the death of her friend. She just would be a little more complete and more able to look at the entire relationship; she would not be stuck on the last incomplete interaction.

Most of us have been taught not to "dwell on the past," and "to let bygones be bygones," and to remember that "what's done is done." All of that well-meaning advice is dangerous because the human mind and heart operate in very specific ways. One of the things that happens naturally is that our brain reviews a relationship and discovers everything that it thinks could have ended differently, better, or more. It is our nature to do this, and we are much better advised to follow our nature than to fight it. The only thing we need to add is what to do with the discoveries after they have been made. As you learn this, you will be able to help your children discover and complete the thoughts and feelings that they recall after someone has died or following any other loss.

Are you beginning to see what we mean by complete and incomplete? It is a crucial distinction, and one you must understand to be able to help your children.

You may never have thought of it this way, but one of the primary purposes of an apology is to create completion of an event that has already occurred. Most of you as parents and guardians are already experts at teaching children to apologize. This lesson is a major part of the socialization of children. It is a lesson that will come in handy as we move from talk to action.

IS INCOMPLETE GRIEF ONLY ABOUT BAD THINGS?

The example we used with the two friends having had an argument might suggest that incomplete grief relates only to bad or negative events between people. That is not true. Incomplete grief exists when there any undelivered communications of an emotional nature. With that definition you can see that anything that has emotional value for your child, positive or negative, can be a part of incomplete communications.

Let's look at an example: A child has received a gift from Grandma. It is a wonderful gift, just what the child wanted. The child plans to write a thank-you note to Grandma. Before that happens, Grandma dies. That child is now left with an undelivered emotional communication about a specific positive event. It is also common to remember something of a general nature that was part of a relationship. For example: "I always meant to tell Grandpa how much I appreciated the time he would sit and talk with me, but I never got around to it."

As a generality, undelivered emotional communications are going to be about things that we wish we had said or done, or about things we wish that we had not said or done. They are also about the things we wish the other person had said or done, or had not said or done. Children are highly sensitive to the sorts of things that fall into those categories. Your job is to help your children discover undelivered emotional communications.

This excerpt from *The Grief Recovery Handbook* further clarifies our point about incomplete communications:

In our three-day personal workshops, we are able to illustrate incompleteness by asking a few questions. On the second day we ask one person if they had any positive thoughts or feelings about one of the other participants. When the answer is yes, we ask what the positive idea was. Usually it is something like, "I admired their courage," or "I liked their openness." We ask, "Did you tell them?" They say no. Then we ask, "What if he or she had died before you told them? Who would be left with the undelivered communication?" They respond, "I would." Then we ask, "If you became incomplete with a stranger in one day, what have you done over a lifetime with family members, friends, and others?"

Incompleteness is not limited to major events. It is an accumulation of undelivered communications, large and small, which have emotional consequence for you. To the best of our knowledge, only the living grieve. It is essential that we complete what is unfinished for us.

Sometimes incompleteness is caused by our actions or non-actions. Other times it is caused by circumstances outside of our control.

One sad story illustrates unfinished emotions caused by circumstances.

A young boy ran across the front yard, hurrying to catch the school bus. As he ran, his mom yelled from the front porch, "Timmy, tuck in your shirt, what will the neighbors think?" Several hours later the police knocked on the mom's front door. Her son Timmy had been killed in a freak accident at the schoolyard.

In addition to the unimaginable pain that the mother was suffering, which last communication do you think this young mother wished had been different? Please do not interpret this question to mean that we are suggesting that if Mom's last interaction with Timmy had been different, she would have felt any less pain. What we are suggesting is that last conversation would definitely fit into the category of things we wish had ended different, better, or more. We rarely ever know which interaction will be our last. It is not abnormal in many of our relationships to have tabled a few topics that we plan to deal with later. This does not have to be procrastination, just planned for later. But, following a death or a divorce, they often become the ingredients of incompleteness.

While death and divorce create obvious areas for incomplete emotions, what about other losses? Often when we look back on difficult relationships with living people, we recognize many things that we wish had been different, better, or more. All too frequently it is the accumulation of these undelivered communications that limits us in other relationships.

Sometimes incompleteness is caused or exaggerated by others. Some people will not allow us to say meaningful things to them. Since we cannot force them to listen to us, we often get trapped with these undelivered communications, both positive and negative. Sometimes we are afraid to say emotionally charged things. Or, we have been waiting for the right moment or circumstances. Sometimes the right time never comes. Or we forget. Or we get sidetracked. And then someone dies. And we are stuck with the undelivered emotional communication.

In short, emotional incompleteness is any undelivered emotional communication. The event or feeling can be recent or from many years ago. Sometimes, we're not sure what we said or did. This can cause feelings of incompleteness. Sometimes we are not sure if the other person heard us, or if they received our communication the way we intended. This also can leave us feeling unfinished.

Please hear this. Being emotionally incomplete does not mean that you are bad. It does not mean that you are defective. It only means that a variety of circumstances have robbed you of the opportunity to be complete.

While those pages were not written specifically for children, they are nonetheless totally accurate for young people who may be even less sure whether their communications were heard and understood. As you think about the idea that unresolved grief is made up of undelivered emotional communications, think about the many clichés we all use about dealing with emotions:

"Keep a stiff upper lip"; "Pull yourself up by the bootstraps"; "Don't burden others with your feelings"; "Be strong."

Those are just a few of the comments that add to children's impression that their emotions are not encouraged, tolerated, or welcome. One of the by-products of not communicating feelings is the inevitable buildup of undelivered communications.

Obviously we are not suggesting that children go up to someone and say anything they feel about that person. Undoubtedly, there are thoughts and feelings that we cannot say directly to living people. But children need a safe place to say those things that would be inappropriate if said directly.

As an example, you may have had an experience where your boss behaved in a less-than-polite manner. You saw it but realized that, if you said anything, you would put your relationship with the boss and your job in jeopardy. So you let it pass. At dinner that night, you might have told your spouse what a jerk your boss had been. To a great degree, you may have become complete with the incident just by verbalizing it safely to your spouse.

We think it's helpful if you can create a similar kind of safety for your children. Sometimes a grandparent, another relative, a teacher, or a coach has been a little gruff or intolerant. Your child, like you at work, may recognize that it would not be safe to tell the adult authority figure how he or she feels in reaction to the words or tone. Just as you used the safety zone of your spouse at the dinner table to tell your story about your boss and to get the incident out of your system, your child might need an interaction with you to do the same.

Remember not to tell your children that they shouldn't feel the way they feel. They already feel that way, just as you did when you told your spouse about your boss. If your spouse was helpful, he or she listened to you, heard you, and let you do what you needed to do. Your spouse didn't tell you that you shouldn't feel that way, or that you should have confronted your boss. We do not mean to suggest that you or your children should remain in an intolerable situation. We are talking about those occasional incidents, which occur for all of us from time to time, at work, at home, and at school.

CHAPTER 10

Helping the Helpers

As we move toward the specific actions that will help you help your children, we want to add a few more general comments that will ensure success. Most people recognize that it can be very difficult to change. You or someone you know may have invested thousands of hours and untold energy in therapy or self-help books and programs.

If we can agree that it is difficult enough to change yourself, even when you want to, then we must allow that it can be very difficult to help someone else change. It is virtually impossible to help someone who doesn't want to be helped. Even when we have better information than they do, if they are not willing, it can be a real problem. You can lead someone to change, but you can't force him or her.

Children need to feel safe if they are going to learn a new, different, or better way to deal with the emotions caused by loss. The following sections will help you create safety so they can follow your lead.

IT'S EASIER TO FILL AN EMPTY CUP

When people sign up to attend one of our seminars or trainings we tell them to "bring an open mind—rent one if you have to, or borrow one from a friend." Even applying as much humor as we can muster, people still want to hold on to the very ideas that are limiting their lives, or, in this case, limiting their children's ability to deal effectively with the loss they have experienced.

We are going to ask you to return again and again to your own assumptions and beliefs, with this question in mind: Is what I am doing going to help or harm my child?

It is very easy to do things just the way we have always done them. But traditions generally signify familiarity, not always value. Using an open mind, look at the traditions you've learned concerning dealing with loss. Naturally, you will want to keep those that are life affirming and beneficial for your children. The rest must be replaced with new methods that can evolve into a new set of traditions that you and your children can use to enhance your lives.

Without going into a long-winded technical explanation of how your brain works, we will say simply that even very willing people sometimes try to cling to old, ineffective ideas. It may be a struggle for you to let different and better information in. Thus the heading of this section: It's Easier to Fill an Empty Cup. Trying to pour something into a full cup can only make a mess.

SCUBA DIVING LESSONS

What does scuba diving have to do with helping children deal with loss? We train professionals and others to use our unique techniques and formats to help grieving people. The first half of the five-day training is devoted to personal completion actions for the people we are training. Recently at the beginning of one of these programs, one of the participants asked why they had to do personal work when they were there to learn how to help others. We answered the question with a question of our own: Would you take scuba diving lessons from someone who had never scuba dived? Certainly not.

Our question to you: Would you ask your children to do something that you are unwilling or unable to do?

Children instinctively look to the adults for guidance. Keeping in mind that your nonverbal actions make up the largest percentage of your communication, the sentence Do as I say, not as I do, becomes both relevant and potentially dangerous.

When your verbal and nonverbal communications match, your children will have an easier time comprehending and following your suggestions.

When they don't match, you run the risk of putting your children in conflict with their correct response to losses. This can have long-term negative impact on their lives.

THE CRITICAL TRANSITION

Infants communicate in the only way they can. They develop a variety of sounds and gestures to indicate what is affecting them and what they want and need. Parents learn to interpret the meaning of different kinds of crying, as well as other verbal and nonverbal communications from their children.

One of life's most critical transitions occurs as children learn to communicate their wants and needs as well as their thoughts and feelings. During this transition children learn verbal language skills and move away from the simpler form of making noises and physical gestures. Information acquired during this critical transition usually has lifelong impact. Of monumental importance is the information observed, adopted, and memorized for dealing with sad, painful, or negative thoughts and feelings.

As always, the timing of this transition is unique to each individual child, and is subject to each child's forming personality and style. Within this framework, the language and gestures of the adults (usually the parents) who are influencing the child play a pivotal role in how well the child assimilates incoming information.

Having worked with thousands of people, all of whom were children at one time, we are distressingly aware of the negative impact of certain kinds of comments people remember from that transition time, comments that have limited their lives. For example, the classic "Big boys don't cry," (re-gendered from the 60's pop song, "Big Girls Don't Cry") sends a powerful message, especially about the transition from infant to more developed levels of communication. The obvious flaw in the comment is the implication that it's never normal and natural to feel sad or to cry when something sad happens, no matter how old you are.

Other examples such as "Grow up," "Be a man," or "Be a little woman," or "Why can't you be more like your big brother or sister?" all contribute to the conflict caused when children are not shown effective ways to communicate sad, painful, or negative feelings. Phrases like "He lost it," or "She broke down" teach children that they are somehow defective when they have the kind of normal feelings that result in tears. Children are smart; they do not want to be perceived as dumb or childish or defective, so they force themselves to accept the idea that it is not safe to communicate their sad or painful feelings anymore. Many years later they come to us at The Grief Recovery Institute with a tragic lament: "My dad died last year, and I have not been able to cry." Is it any wonder?

Just twelve hours before the writing of this page, Russell was on a radio interview with call-ins from a large regional audience. An elderly woman called in and made the following comment: "I am seventy years old, and I have never felt sad." It took Russell just a few moments to determine that the woman had grown up in an environment that was totally intolerant of most human emotions, especially sad ones. Both her mother and father were from the "old country" and did not believe that those kinds of feelings were good for people. In fact, this woman had not participated very much at all in any feelings, even happy ones. Russell took a risk and asked her the following question, "Does it make you sad that you have never been able to feel sad?" There was a long silence before she answered, in a small voice, "Yes." Russell used her response to help her see that her heart had not been totally crushed by those ideas she had learned nearly seventy years ago. Her closing comment was that she felt some hope for herself as the result of listening to Russell and realizing that she could feel some sadness after all.

BOUNDLESS CAPACITY

The capacity for human emotion is boundless. Just witness infants who squeal with delight at the tiniest pleasures yet moments later cry with agony. It is not just the unlimited capacity for emotion that we are concerned with, but also the ability to communicate an individual feeling, either positive or negative, and then move on to the next one. Sort of like that Midwestern joke about the ever-changing weather: If you don't like the weather, just wait a minute. If you don't like the feeling you're having, another will be along in a moment, if you let it. Infants don't have a choice, so they bounce along from feeling to feeling, happy to sad to happy to neutral, in a perfectly accurate and emotionally honest response to themselves and to the world around them.

> It is the expression of the current feeling that
> allows the expression of the next feeling.

Restricting the expression of naturally occurring emotions can cause a buildup that, in turn, can lead to an explosion. Think of a steam kettle with a little water inside, a flame roaring underneath, but no spout to let out the buildup of energy. We all need the ability to relieve properly some of the pressures that build up inside. The next chapter illustrates the long-term buildup of incomplete emotions that can result when little steam kettles are not correctly tended.

Without oversimplifying, we think it's reasonable to say that the two most basic human emotions are happiness and sadness. We think they are of equal importance and must be voiced, as needed, to allow movement to the next range of feelings that accompany the events in each of our lives. Holding on to feelings, particularly sad, painful, or negative ones, consumes much more energy and can have many more negative consequence than expressing them immediately and appropriately when they occur.

What is appropriate obviously bears a relationship to both the age and maturity of children. Very young children who are just learning how to talk will retain many of their infantlike expressions. On the other hand, children further along in this transition will be better able to express their feelings verbally, along with tears, laughter, and other nonverbal communication.

DELICATE STROKES

We tend to limit the expression of the sad side of human feelings by using an overly broad brush when we would be better off using more delicate strokes. At The Grief Recovery Institute, even when working with adults, we say, "Talk while you cry," which helps people communicate specifically what they are feeling, as opposed to expressing only the generality of sadness implied by their tears. When helping children, we do the same and gently encourage them to add words to the feelings they are trying to communicate. The danger is that if we tell them to stop crying or imply in any way that there is something wrong or bad about their crying, we will break the essential bond of trust and safety that we have created for them.

The transition from infantlike communication to verbal communication skills can be very frustrating for parents and other guardians. Some children sail through without a hitch, while others stop and start, two steps forward and one step back. It is essential that you remember that this transition is specific to each child. Again, those of you with more than one child will already have observed the often-surprising differences in the pacing of each child's transition. We know that this transition can be difficult and frustrating at times, but we would caution you to be as patient as possible.

> Your words and deeds can have powerful
> impact on your children; it is essential that they
> not feel judged as they make this passage.

IF YOUR KIDS ARE OLDER, DO NOT DESPAIR

If your child or children have already gone through that basic transition we have been referring to, please do not despair. Children are
more flexible than adults. They have had less practice than adults at
habituating incorrect ideas. Given different and better information,
they will adapt fairly quickly. Russell remembers coming back to the
office after having spoken to a seventh-grade and then an eighth-
grade class. He told John that the seventh-graders seemed more
emotionally available than the eighth-graders. It took a little longer,
but the eighth graders started to participate also.

DO WE KNOW ENOUGH YET?

Everything you have read up until now serves to illustrate the fact
that a problem exists. It is almost impossible to introduce the
actions that lead to completion of the pain caused by loss without
first examining the foundation beliefs that you currently hold.

A cut may become infected. If not treated, it can get worse. If left
unattended, there can be disastrous consequences. Here is a parallel:
Unresolved grief is cumulative and cumulatively negative. Unresolved grief does not go away by the counterproductive nonactions
of trying not to feel bad or by replacing the loss or by grieving alone
or by being strong or by keeping busy or by the passage of time.

In taking an in-depth look at the six major myths that dictate
most of our reactions to loss, you no doubt have begun to recognize
that you must make a few changes in your own perceptions so you
can help the children.

Understanding the pitfalls contained in the six myths is the first
essential component of helping your children. While you now
know a lot more than you did, next we will look at another very
important element that contributes to the accumulation of unexpressed emotions.

CHAPTER 11

Short-Term Energy-Relieving Behaviors (S.T.E.R.B.s)

Have you ever known anyone who had a very powerful negative attachment to something that had happened a long time ago? We are referring to a person who tells the same tale of woe, over and over again. And no matter how many times the story is told, it contains the same tremendous emphasis, pain, and anguish. Years after the event, it is the same story, still attached to the same feelings, using up huge amounts of energy.

Did you ever wonder how much energy that person was consuming just to hang on to that story and to retell it over and over?

When we talk about energy, we are not talking in mystical or magical terms. We are talking in everyday terms about the amount of tangible human energy it might take to hold on to and relive a painful story over and over.

Did you ever wonder if the regeneration of painful memories might begin to take a toll on someone's health? Could this be the kind of thing that creates or contributes to ulcers, high blood pressure, headaches, stomach problems, backaches, and an almost unlimited number of other illnesses?

In fact, we believe that there is a direct correlation between undelivered emotional communications and a whole host of medical conditions. Even though children's bodies may be younger and seem more resilient, they are just as liable to suffer the negative physical consequences that accumulate in adults who do not deal effectively with their emotions. Since unresolved grief is cumulative, it may take some time before there are observable symptoms.

Ask yourself this question, Does it take more energy to hold on to something or to let it go? Just for fun, make a fist and really clench it hard, so no one can break into your hand. Observe how much energy it takes to make that fist and hold it clenched. Keep clenching it for a while. You might even begin to feel a little sore. Eventually your arm and hand will tire, and you'll have to unclench.

Now, take the same test, only this time clench your stomach and hold it. See how long it takes before that doesn't feel very good. Imagine that your tummy felt that way all the time, because you were holding on to something that had happened sometime in the past, or holding on to the fear of something that might happen in the future. Imagine how much energy is consumed either by your fist or tummy, trying to hold on.

One thing should be fairly obvious; it takes a whole lot more energy to hold on to something than to let it go. Well-meaning friends tell us that we should let go and move on. Most people would do so if they knew how.

The questions then become, Why do we hold on to painful events and memories so tightly? and Why is it so difficult to let go?

The answers are partially buried inside some of the six myths that we discussed in part one of this book. When we have been socialized to believe an idea like don't feel bad, it causes us to be in conflict with our own nature. Consequently, when something does happen and we do feel bad, but we believe we're not supposed to, we push the feelings down or away. In addition, we may replace the loss with a substance of some kind, or with a parallel relationship. A new relationship does not complete the old one, it just buries it out of our consciousness. The feelings attached to the original relationship have not been dealt with, and in its place there is a different activity to distract us from the feelings we are not supposed to be having.

Consider the myth that advises us to keep busy. Keeping busy certainly generates and consumes a lot of energy, but the emotional energy caused by the painful event is not addressed; it is simply ignored while we concentrate on other activities. By the time you have taken other actions, you may not even remember the original loss.

DO YOU KNOW WHERE YOUR CHILD'S ENERGY IS?

Earlier we suggested that one of the reasons you are reading this book may be because you have observed some behavior in your child that you sensed has been caused by a loss, either recent or in the past. There is a very strong possibility that what you have been noticing in your child is what we call a short-term energy-relieving behavior or s.t.e.r.b.

Grief, the normal and natural reaction to loss, can produce measurable energy in a child. When the normal flow of that energy is sidetracked by advice such as, "Don't feel bad. Have a cookie; you'll feel better," the energy itself still exists but has merely been shunted aside. Incidentally, the improper use of food in a misguided attempt to deal with feelings is the biggest single s.t.e.r.b. in North America today.

Once again we refer to the little girl we mentioned in chapter two, who was given cookies when she cried about how the other kids had been mean to her on the preschool playground. She is told to eat the cookies and she will feel better. As the result of eating the cookies, the child feels different, but not better. Following a series of events like this, a child will identify different as better, because that is how the adults have defined it. Consuming a substance in response to an emotional event often becomes the cornerstone of an unconscious philosophy that can have lifelong negative consequences. It also becomes the beginning of an unintentional and incorrect habit for dealing with grief and loss.

Fast forward: ten years later at the junior high school playground, following an emotionally painful interaction with a classmate, the local marijuana pusher says, "Feeling bad? Try this, you'll feel better." Where have we heard this line before? Painful emotions? Consume a substance and you'll feel better. And we don't have to rely on illegal narcotics and a shady character to make our point. For example: A teenager is feeling depressed after a romantic breakup. She tells her mom and dad, who do not know how to deal with those normal feelings of grief, so they take her to their local doctor or therapist. Often, as the result of one visit, a psychopharmaceutical drug is prescribed; "Don't feel bad. Take these pills; you'll feel better." Now, in a legitimate setting, a legal substance is prescribed to deal with the feelings, with the same potentially damaging results.

Our society has been focused on the dangers of alcohol and drugs, while at the same time it's promoting substances for dealing with feelings. This is an incredibly mixed message. Do you think there may be any connection between the fact that we are taught to medicate our feelings with food and other substances and the fact that there is an average of two hundred eighty thousand obesity-related deaths each year in the United States? Could this be the direct result of socialization that relies upon such concepts as, "Don't feel bad. Have a cookie; you'll feel better?"

We have to wonder: What happens to the sad, painful, or negative feelings? Do they go away? What happens to the tangible energy produced by those sad emotions? Does that energy just disappear, never to return? Or is this energy merely covered up, to come back, along with the attendant sad feelings, and haunt the child later? Do those feelings become a ticking time bomb buried just under the surface, waiting for the "wrong" stimulus to trigger an explosion? You bet your life—or, more accurately, you are betting your child's life. You wouldn't do this intentionally to your child, but let's make sure you have better ideas and better ways to communicate about loss with your children.

SHORT-TERM RELIEF DOESN'T WORK

In the last chapter we referred briefly to the idea of the buildup of pressure inside a steam kettle. In *The Grief Recovery Handbook* we wrote extensively about the short-term energy relief using the image of the steam kettle to punctuate the message. Over the years, we have had so many positive comments about the clarity of that explanation that we are reprinting that section here, with slight modifications to reflect our focus on children.

Imagine a steam kettle. The kettle is filled with water. The flame under the kettle is turned up high. Normally, as the water heats and boils, the steam generated by the heat releases through the spout. Most kettles are fitted with a whistle to notify us when they have reached the boiling point. Imagine that same steam kettle filled with water, with a high flame burning below, but now there is a cork jammed into the spout. Imagine the pressure that builds up inside that kettle when the spout cannot release the built-up energy. The cork represents a lifetime of misinformation that causes us to believe that we are not supposed to talk about sad, painful, or negative emotions.

A healthy steam kettle releases energy immediately as it builds up. When children are told "Don't feel bad," and "If you're going to cry, go to your room," the energy—or steam—stays inside. The major myth that time will heal is laughable if you think of the steam kettle. Time will only move the steam kettle in our illustration closer to an explosion.

Now think of your child as the steam kettle. As the pressure builds up inside childrens' personal steam kettles, they automatically seek relief. This is when they may start participating in short-term energy-relieving behaviors. There are three major problems with s.t.e.r.b.s. The first is that they work, or more accurately, they appear to work. They create an illusion by causing the child to bypass or bury emotions. The second problem with short-term relievers is that they are short-term. They do not last, and they do not deal with the presenting emotional issue. And lastly, they do nothing to remove the cork that is jammed in the spout. In fact, most children don't even realize that there is a cork in the spout.

Eventually the child's little steam kettle is overloaded and the s.t.e.r.b.s. no longer create the illusion of well-being. Imagine what might happen if a major loss event, a death or divorce, was added to this collection of unresolved emotions. It might put such a strain on the corked kettle as to actually cause an explosion.

Some emotional explosions are huge and make national head-lines, like the tragedies at Columbine High School and the junior high school in Jonesboro, Arkansas. Most are much smaller. Here is an unfair question. Have you ever had an emotional explosion larger than circumstances called for? Sadly, we know that most of you have to say yes. Even something as simple as coming home from a tough day at the office and yelling at the person who left a roller skate in the hallway is an example of such an explosion. Over time, we develop the habit of putting a cork in our own per-sonal steam kettles. We bottle up our own feelings because we have been taught to do so. We're good little students, so we do as we're told.

Our children tend to do what we do. Remember, they have been copying us since day one. So, whether the message is direct, in the form of a comment like, "Don't feel bad," or indirect, through a child observing a parent's behavior, the result is the same. Another generation has learned to deal with feelings indirectly instead of directly.

The ideas in this book will allow you to educate your children differently. In effect, you will be able to help them remove the cork and make appropriate and timely expression of all of their emo-tions. In addition, the actions of this book will help you guide your children in removing the cork that may have been there for a while. They will then be able to deal more effectively with the emotions associated with loss. By looking so carefully at the myths and other incorrect ideas, you can now help your children replace them with more accurate ideas about dealing with sad, painful, and negative emotions.

A simple analogy: If your yard is full of weeds and you cut the weeds, you will create short-term relief. Short-term, because the weeds *will* grow back. Or, you can *pull* the weeds and eliminate the problem altogether. We want your children to have an automatic predisposition to *pulling* the weeds. That way they won't have to have lifelong battles with substances and other emotional substitutes that are caused by the illusion of well-being built into short-term energy relief.

Food, drugs, and alcohol are the most obvious examples of s.t.e.r.b.s. But there are many others. Some of them are more relevant to children, but here is a list of the most typical ones.

> Anger
> Fantasy (video games, computers, movies, TV, books)
> Isolation
> Sex
> Exercise
> Shopping (humorously called retail therapy)

Following major losses of any kind, children are likely to generate tremendous emotions. Their bodies will naturally try to deflect the excess energy, which can evidence itself in any number of ways. Most of you are familiar with the phrase "acting out," which often represents the overwhelming excess of energy caused by loss. Sadly, our schools and teachers are not as well equipped as they could be to recognize and distinguish between a real behavioral issue and the impact of grief or unresolved grief on young people.

Depending on age and other circumstances, the vast majority of young people begin their involvement with drugs and alcohol soon after a major loss experience. We have already gone to great lengths to explain how early the use of food is incorrectly applied to emotions.

We cannot encourage you strongly enough to take these ideas seriously. And, in the matter of short-term energy relief, we would strongly suggest that you look closely at your own behaviors, as well as those of your children.

RECAPPING PART TWO

We have shown you the ideas and actions that keep children from becoming emotionally complete. You now can see more clearly how the six widely accepted myths lead children away from completion. You can also see that s.t.e.r.b.s, which have limited value, wind up contributing more to the problem of unresolved grief than to the goal, which is completion.

With those obstacles out of the way, we can now start talking about the ideas and actions that lead children to completion of the pain and unfinished emotions attached to the people, animals, places, and things that are important to them.

PART THREE

The Path to Completion

WHAT IS COMPLETION?

Completion is the action of discovering and communicating, directly or indirectly, the undelivered emotions which attach to any relationship that changes or ends.

Deaths, divorces, and other major changes leave children with undelivered emotional communications about things that did and did not happen within their relationship with the person, animal, place, or object. Those undelivered communications are the components that make up unresolved grief.

It is virtually impossible for a relationship to end without containing some unfinished or undelivered emotions. Try as we might, it is exceedingly difficult to sustain a condition of absolute completeness with anyone or anything. Our thoughts, feelings, and opinions are constantly changing; therefore, our relationships are also continuously undergoing minor or major shifts.

In order to determine which elements of a relationship are emotionally incomplete, we must help our children review the relationship. The relationship review is the first major action that will lead to the communication of undelivered emotions.

CHAPTER 12

The Relationship Review

RELATIONSHIP REVIEWS HAPPEN AUTOMATICALLY

When a major change in the circumstances of our relationship with someone or something occurs, we automatically review the relationship. A death, a divorce, or even a move will cause us to look at every aspect of that relationship right up to the present time. It is impossible *not* to review a relationship when a loss has occurred. People might tell you, "Don't think about it," or "Don't dwell on it," but not only is that bad advice, it's also impossible to follow. It is in the time immediately following a loss that memories of the relationship are most accessible, accurate, and intense.

A relationship review is an automatic and natural response to loss for everyone. The only potential differences between adults and children in this matter are likely to relate to the length of time a relationship existed and the adult's more developed ability to communicate thoughts and feelings about it.

Some children are consciously aware of the review while it is happening, and some are not. Immediately following a loss, a death in particular, it is common for a family to sit around and talk about the person who died. Each family member will tell the stories that most identify their personal relationship with that person. Some of the stories will be shared by the entire family. Some will be unique to each individual. The children's stories will always relate to their personal interactions with the person who died. This kind of family conversation is a relationship review. It is normal, natural, and healthy, and should be encouraged, for it will assist you in helping the children. We strongly recommend that you encourage your children to participate in this kind of family conversation. You may be surprised at the different kinds of memories that have affected them, either positively or negatively, in that relationship.

While family conversations about a loved one who has died are normal, they should not be confused with completion. Completion usually requires additional actions. However, stay alert to the idea that the memories brought up to the surface following a death can be very helpful in discovering things that a child wishes had been different, better, or more, and may also include unrealized hopes, dreams, and expectations about the future. Those discoveries can be used to complete what is emotionally unfinished.

WHO GOES FIRST?

We are ready to move from talking about recovery, to outlining the actions that will help your children complete the pain caused by their losses. But just before we make that move, we must make something very clear. You might recall at the beginning of this book we said: Recovery from grief or loss is achieved by a series of small and correct action choices made by the griever.

The most important part of that statement is that the choices are *made by the griever*, who, in this case, is the child. The choices are not made by the parent or guardian. We do not believe you can force recovery on a child. We have seen too many bad results when well-meaning people try to force grievers, either children or adults, to do what *they* think is best for the griever.

If you consider the evolution of this book, you will note that John and Russell and Leslie went first. Each of them shared some of the loss experiences in their lives, and the ideas and philosophies they learned about dealing with losses of different kinds. In going first they have established the idea that if they can do it, you can, too. You also need to go first, so your children will feel safe enough to make the choices that you will illustrate for them. Remember, we asked if you would take scuba diving lessons from someone who has never scuba dived.

We also want to make clear the fact that you, as parent or guardian, know your children best. Those of you with more than one child will recognize that each child has his or her own unique communication skills. You will know the ones who need more help identifying their thoughts and feelings. We are going to give you some universal guidelines that have proven successful for all losses and for children of all ages. You are going to have to use your awareness to identify your childrens' individual maturity, comprehension, and communication skills.

Children have their own individual and unique relationships. You must help them discover what is emotionally unfinished for them in their relationship with the person or animal who died. With losses other than death, you must still be careful to remember that these actions are about *their relationships*, not yours. The biggest possible danger is for you to plant *your* interpretation of their relationships into their minds. Since children are young and impressionable, they might be inclined to take on what you tell them and not discover what is true for them.

PICK THE FRUIT WHEN IT'S RIPE

There is no perfect, or even correct, time and place to begin to help your child review a relationship. We can't presume that each of you is at the exact same starting place with your child. Some of you may have naturally been encouraging your children to talk about the emotions they are experiencing about the relationship under review. Others of you may have used old ideas, such as, What's done is done, or Just move on. We are not here to judge you for what you've done or not done until now; instead, we want to provide helpful guidelines.

Most children will take their cues from you. If you talk openly and honestly about your relationship, they will, too. If you allow yourself the normal, wide range of emotions that attach to loss, so will your children. If you don't apologize for having human emotions, they won't get the message that these feelings are abnormal or defective.

Asking direct, interrogative questions usually discourages children from talking about how they feel. If you ask your child, "Have you been thinking about Grandma?" you are liable to get no for an answer. And you may have derailed discussion of the topic, at least for the moment.

In simplest terms, children say no because they are afraid of being judged or criticized for having the feelings they are having. Remember our first myth: don't feel bad. In and of itself, that comment implies that we are somehow defective if we feel bad at all, or if the feeling continues for more than a moment. We are taught that we should not feel bad, and that if and when we do, we must go to our room. We memorize and habituate the idea that it is not safe to feel bad, and, even worse, it is not safe to feel bad in front of others.

Often the best way to find out what is going on with your child is to talk about yourself first. Imagine that you say to your child, "Since Grandma died, I've really been thinking a lot about all the visits we made to her house. I remember the time she forgot we were coming, and she was so upset because she hadn't made any cakes or pies." An opening sentence like that will automatically encourage your child to think about things that happened in his or her relationship with Grandma.

Because of our culture's predisposition to minimize painful feelings, most children have learned that when they tell the truth about sad feelings, they will then be judged, criticized, or belittled. We are not saying that you have done those things. But negation of emotional pain is all-pervasive in our society, so even if your children were not treated that way by you, they are still likely to have been affected by the world around them.

When you ask a child how he or she feels or what he or she is thinking or remembering after Grandma dies, he or she may not be willing to reveal the truth for fear of being told that these thoughts or feelings are not okay. However, if you will remember to go first, and talk about your own feelings, you will eliminate the child's fear that you will be judgmental.

It is also a good idea to include an emotional word or phrase in the things you say. If you use emotional words, you are indicating to your children that it will be safe for them to talk about how they feel. Let's go back to our original comment: Since grandma died, I've really been thinking a lot about all the visits we made to her house. I remember the time she forgot we were coming, and she was so upset because she hadn't made any cakes or pies. It would be very helpful to add, "I've really been missing her and feeling sad."

Remember, you are the guide.

CHAPTER 13

Real-Life Examples

OUT OF THE MOUTHS OF BABES—
GOOD-BYE, MR. HAMSTER

The best illustrations we can give you are from real-life stories. The following story happens to concern the death of a pet. Up to a certain age, most children respond perfectly to loss experiences. They will display any and all emotions. They will review and complete what is unfinished for them. And then they will move on. The age at which they begin to give up this natural ability differs, ranging anywhere from about three to seven. Notice we said "give up this natural ability." We want to emphasize that point. Children do things correctly until they are taught otherwise. We received a call from a friend of ours a few years ago. Her four-year-old son's pet hamster had just died, and our friend was very anxious that she be able to help him deal with the feelings he would have in response to the death of his pet.

We immediately told her that she probably wouldn't have to do too much; since he was four, he would probably get it right all by himself. At four, he had not been overexposed to literature and films that constantly send faulty messages about dealing with loss. We suggested that she just watch him and see what he did; if there were any problems, she could call us back immediately. An hour later she called back, somewhat amazed. She asked how we knew. We said, again, that at age four telling the truth is still both possible and likely. She told us that she had been able to watch her son in his room, as the door to his room was open just enough that she could see him. He stood in front of the cage, looking in at the dead hamster, and with tears in his eyes he said, in four-year-old fashion, "Mr. Hamster, you were a good hamster. I'm sorry for the times I didn't clean your cage. I was mad the time you bit me, but that's okay. I wish that you didn't have to get sick and die. I wanted to play with you more. I loved you, and I know that you loved me. Good-bye, Mr. Hamster." And off he went.

That is not the end of the story, only the beginning. Later, the little boy, his mom, his dad, and his seven-year-old sister had a little ceremony. Mr. Hamster was buried in the backyard, in a shoebox, along with some pictures that he and his sister had made. Mom encouraged him to repeat the things he had said in his room, and to say "Good-bye, Mr. Hamster." From time to time over the next several weeks, the little boy would go out in the backyard and talk to Mr. Hamster, and he always remembered to say good-bye. Two months later, the little boy went to his mom and said that he thought he might like to have another hamster now. She thought that was a splendid idea. The very first thing the little boy did when he got his new pet hamster was to tell the new hamster all about Mr. Hamster, who had died, and to say that he hoped that they could become good friends, too, just as he'd been with Mr. Hamster.

> Kids are awfully good if you let them be. Sometimes
> we think it's better to follow them than to lead them.

The little boy in the story had done a relationship review without having been told to do so. It is natural because the awareness of the death automatically prompts the review. This little boy, at four years old, had not lost the instinctive reaction of reviewing and communicating his thoughts and feelings about his pet.

Since he was only four, his memories and the way he talked about them were an accurate representation of his relationship with Mr. Hamster. Older children might have had a wider variety of memories, as well as more sophisticated ways of talking about them.

ALL GRIEF IS EXPERIENCED AT 100 PERCENT

We are starting with the death of a pet, because it is very often the first major loss that affects a child. But, we never compare losses. Please do not interpret this to mean that we put a higher value on the death of a pet than on the death of a parent, a grandparent, or any other person. All losses are experienced at 100 percent, and each loss carries with it a level of intensity based on the uniqueness of the relationship.

THE DEATH OF A PET

The death of a pet can cause considerable emotional pain for a child. You will recall that the death of John's dog, Peggy, established the litany of misinformation that affected much of the rest of John's life. It was from that experience that the ideas "don't feel bad," and "replace the loss" were etched into John's consciousness.

If the death of your child's pet is the reason you are reading this book, then there is a strong possibility that you also had a relationship with the pet. If the pet was a dog or cat, it is likely that you also had an emotional attachment to the animal. On the other hand, if the pet was something like a hamster, a snake, or some other caged animal that lived in your child's room, your relationship with that pet might have been minimal.

Since we are focused on helping children, the details of your relationship with the pet are not the key issue. But your own memory of parallel events in your childhood can be helpful. When you were a child, did you ever experience the death of a pet? If so, try to remember how you felt. Doing so will give you a sense of connection to what your child is experiencing. While there is no universal set of feelings when a pet dies, most children are strongly affected.

If you did not experience the death of a pet when you were a child, try to remember any sad experiences that you might have had. Even the loss of a prized possession, like a teddy bear or a baseball bat, can be extremely upsetting to a child. Any memories you have of sad or painful events will help you access your own emotions, and will help you guide your children to their own emotions and to completion of the pain.

Since relationships with animals tend to be unconditional, children, like adults, form very intense bonds with their pets. Children will often tell their pets all of their thoughts and feelings and frustrations. The pet becomes a trusted confidant—unconditional and totally trustworthy.

The death of a pet, especially when it is the first death of someone close to the child, introduces the painful reality that something can go away and not come back. The permanence of death becomes a painful fact as opposed to a theoretical possibility. As a parent or guardian, you will be wise to remember that in addition to the actual death, this incident may be hammering home to your child the finality of death. Recognize that huge emotional responses to an event of this magnitude are very normal for children.

Everything we have been teaching so far becomes relevant now. Here is a question: Is it true that all living things must eventually die? The answer is yes. Here is another question: Is it helpful, in response to the death of the child's pet, to address the child's emotions with the fact that all living things must eventually die? The answer is no. The death of the pet might provoke the most intense emotional response your child has had to anything in his or her young life. The fact that all things eventually die is intellectually accurate but not emotionally helpful to the child, at least not at this time.

In fact, the comment "All living things must eventually die" is really a shortened version of one of the myths: Don't feel bad, all things are going to die. Can you see, when we put it in that context, how unhelpful it might be to your child? It is important that you accept the fact that the child's emotional reaction to the death of his pet, who the child perceived to be his best friend, is normal and natural. Any comment or implication that he shouldn't feel bad puts him in conflict with his own natural response. Your very first task is to avoid the knee-jerk reaction of saying, "Don't feel bad." We talked earlier in the book about the natural desire to comfort our loved ones. Telling a child not to feel the way he or she feels is not comforting; it is confusing.

You may wonder if there ever is an appropriate time and place to talk about the fact that all creatures die. In fact, many of you probably have already had that kind of conversation with your children. For those of you who haven't, we will talk about that later in the book in chapter thirty, *The "D" Word*.

However, it is very typical for all grievers, children and adults, to ask the age-old question "Why?" when emotionally affected by a death. Please recognize that it is not really an intellectual question, but rather an emotional lament. Remember that none of us ever wants our loved ones to die, and it is never okay when it happens. Before you rush in with a big intellectual definition of death, make sure that you're responding to the real question.

It might be much more helpful and supportive, in response to your child's Why?, for you to say, "I don't know why, honey, but it sure is *sad* that Fido died, isn't it?" Do you see how that answer acknowledges the question, and at the same time brings it back to the feeling?

If your child really wants a scientific explanation, he or she will make that very clear. If you automatically move toward the intellectual explanation, this was likely the response you heard when you were a child. You may be unwittingly repeating something that was said to you a long time ago.

The death of a pet is, first and foremost, an emotional event for your children. Their primary reaction will be to the painful feelings they are experiencing. Your job is to acknowledge and assist them in recognizing, accepting, and dealing with the truth of those feelings. If you shift them away from their emotions and toward their intellect, you will have done them a disservice. If you do this, you may have experienced this as a child, or you may be afraid that you do not know how to help your children with their feelings.

Let's get back to you. If you had a pet die when you were a child, then you probably remember how much it affected you. This event can be a powerful teaching opportunity. Imagine if you could say to your children, "I can see that you are heartbroken at the death of Fido. I remember when my dog died when I was a child. I was so sad. I missed him so much." Can you see how those few simple sentences might help them feel safe enough to talk about how sad they might be feeling? Those comments create safety so children can tell the emotional truth. Notice that these comments make no movement toward any intellectual ideas.

You cannot shield yourself or your children from losses in the future. But, by creating the safety to communicate the normal and painful feelings about loss, you are giving your children a solid foundation for dealing with the painful events that will occur from time to time throughout all of their lives.

What we have illustrated is only a beginning. Obviously, just making one supportive emotional statement will not end your children's grief, nor will it make them complete with any unfinished emotions attached to the relationship. The initial reaction to the death of the pet, while powerful, is a preamble to the feelings that follow. Here again is our definition of grief: the conflicting feelings caused by a change or an end in a familiar pattern of behavior.

In the days and weeks following the death of a pet, there are constant reminders of the pet's absence—the empty food bowl, the missing sound of little feet, no walks outside—each of which contribute to the painful reality that the child's beloved companion is no longer there. The feelings of loss are not limited to the day of the death. Feelings don't end following a funeral, burial, or memorial service. Feelings continue.

Even if it has been some time since your children's pet died, it is never too late to introduce the idea of a relationship review. As you think about the death of your children's pet, you will automatically remember parallel or similar incidents in your own childhood. You can use your memories to guide you to recognizing what your children are experiencing. If you never had a pet, we will share with you some of the most common areas that are part of a relationship review with an animal. If you can accept the idea that the death itself creates a naturally occurring review process, it makes no sense not to make positive use of the memories that are already popping up in your children's minds and hearts.

RANDOM MEMORIES

While the review of relationships with people or events is automatic, it does not always happen in an orderly, sequential fashion. It is more typical that children will have random memories of a wide variety of events. Some of the events may evoke fond memories, while some may be bitter or painful. The mixed emotions associated with the relationship with an animal or person who died can be very confusing for children. Part of your job is to help your children understand that it is normal to have a mixture of positive and negative memories and emotions.

When you are helping your children review a relationship, don't be concerned if they bounce around as they recall the incidents that pop into their minds. It is more important that you recognize the fact that they have remembered something, not exactly when it happened. There will be time later to put the events in sequence, if necessary.

CHAPTER 14

Helping Your Child Review the Relationship

All relationships begin at the beginning, but that does not mean that all relationships begin at the first meeting. Some families plan extensively for the adoption of a pet. Often the children are included in the search and selection process. During all of the activity leading up to the selection, children's hearts and minds will start to build up hopes and expectations about what the relationship with the pet will be like. In this sense, their relationship to the pet begin with the idea of having a pet. When helping your children to remember their relationship with the pet, don't leave this part out. Sometimes there are major emotions attached to how the search process unfolded.

Very often children have to lobby for a long time to get their parents to grant permission to have a pet. Children have to make loyal oaths to commit to the care and feeding of the pets. In some cases, this may be another hidden area of emotional content after a pet dies, one that reminds children just how much they wanted the pet.

A moment that often contains an unlimited amount of emotional energy for your children—and maybe for you, too—is the moment when they saw the animal for the first time. This moment can be especially critical since animals have the capacity to elicit tremendous responses with their eyes and body language. It is often impossible to resist those precious moments of the first interactions with an animal.

Over the years we have received thousands of phone calls from grieving pet owners, adults and children alike. Many of them cry when they begin to tell us about what happened to their beloved pet. At the appropriate time we ask them if they can remember the very first time they actually saw the pet. Most callers say yes, and as they start talking about that moment, their voices change. We can sense a difference in the caller, even through the telephone. The memories of that first meeting are so potent that they produce smiles even through the tears of sadness. It is very likely that your children will have very strong memories of their first meeting with the pet. Make sure you bring this up. It can be an emotional gold mine.

There may be instances when your children do not have vivid conscious memories of meeting the pet. Remember John's dog, Peggy, who was six years old when John was born? He can't really remember the moment he met Peggy; she was just always there. When she died, John was devastated because she was no longer there.

The early days of a relationship with a pet are often filled with a variety of both positive and negative events, both of which can elicit powerful emotional memories. Often baby animals, just removed from their mothers and littermates, may cry all night. Even though this is normal, your children might have become afraid that you would take the pet away if it continued to keep the family up all night. Housebreaking, teething, scratching furniture, and any other kind of disturbances may evoke feelings for both children and parents.

The process of training, especially with dogs, may be filled with emotionally charged incidents. Some of the events are funny, some are frustrating or anger provoking, and some are even dangerous. Decisions about whether your pets are housebound or are free to roam outside may play a strong part in children's emotions about the loss. It is not uncommon for pets to be hurt or killed by vehicles on the road. Some cats and smaller dogs are often killed by coyotes or other wild animals, and even others meet untimely ends as the result of poison. If your children's pet died as the result of being an outdoor pet, there may be some feelings about those circumstances.

As we walk you through all of the areas that are bound to contain emotions for your children, do not make the mistake of judging them for having feelings about any of the things we are discussing here. Your primary task is to create safety for expression of any and all of the feelings your children attach to their relationship with the pet.

Sometimes pets get ill, get hurt, and even die as the result of following some of their wilder instincts. We often forget that their ancestors were not housebroken and did not know that they were not supposed to go tearing into pails of garbage. Memories of incidents of this type will also create feelings in your children. Don't try to bypass memories that you think might be too painful for children. It is far better to allow them to feel all of the emotions that are a part of their relationship with the pet.

Your job is to help them remember and, if necessary, to help them discover the feelings that best describe their reaction when the event occurred. For example, one of our friends related an incident when their dog got into the garbage and cut her paw on the sharp edge of a tin can. Our friend told us how she reminded her daughter how scary that had been and what it was like racing to the veterinarian and being so worried about the pet. After the visit to the vet, when everything had turned out okay, she recalled how relieved the child had been. Did you notice how the mom used the words, *scary, worried,* and *relieved?* All of these are emotional words.

SLEEPING IN THE BED, OR NOT

Many children want the company of their animals in bed with them at night. Some parents allow this, some don't. In those circumstances when it has been allowed, the children may have some additional emotional responses at bedtime. Remember our definition: . . . change or end in a familiar pattern of behavior. In some cases you may have to cuddle up with your children at bedtime for a while, as they adapt to the new reality of life without their pet.

All of the routines associated with the children's relationship with the pet will produce emotions. Feeding, walking, playing, and grooming are just a few of the daily interactions that bonded the children and the pet. Built up over long periods of children's lives, the end of those routines can be powerful and painful.

MINDING THE STEAM KETTLE

As you continue to help your children review the relationship with their pet, you will eventually arrive at the moment when the pet died or got lost or ran away. When there has been a long illness, the details of the onset, the treatment, and even the medications may be very important to your children. Sometimes a particular visit to the vet, which determined a cancer or other condition, will be remembered with great intensity. Don't overlook any of these areas. You want to help children discover and communicate about anything that produces feelings.

Like the steam kettle we talked about earlier, children's bodies will want to release some of the energy that is being generated by the death of the pet. In the upcoming pages, we are going to explain and list the areas that are most likely to contain emotional energy.

Do not make the tragic mistake of telling your children that they should not be thinking, feeling, or expressing the emotions they are having in response to the death of their pet. If you do, you run the risk of instilling an idea that will eventually cause your children to explode or implode.

There is no way to predict which events children will remember in relationship to the pet that died. If you have more than one child, each one may place emotional value on different events. And you, yourself, may have an entirely different set of memories about that pet. The key here is to remember that it is the specific memories and emotions unique to each individual child that must be discovered *by that child*. You must be very careful not to plant ideas based on your relationship and memories about the pet so that you don't falsely influence your children. This can be a little tricky; just try to stay alert.

CHAPTER 15

The Emotional Energy Checklist

CHILDREN AND THEIR PETS:
REVIEWING EVENTS AND EMOTIONS

All losses create emotional energy. All losses cause children to review the relationship that has just ended or changed. Contained within the naturally occurring review are the things the children remember. Some of the memories concern happy events, and some of the memories concern sad ones. Some of the memories do not contain much emotion; they are simply things the children remember. While the death of a pet provokes a natural process of review, the same review happens in response to the deaths of people, moves, divorce, graduations, and other losses.

Unlike the little boy whose hamster died, not all children will have all of their memories contained neatly in a couple of sentences. And all of the memories may not come to children at one time. Again, older children, with longer and more intense relationships with pets, and with more complex communication skills, may have more to discover and more to say.

We have prepared a list of the events and ideas that we talked about in the last section. The list represents the areas that are most liable to produce both positive and negative emotional memories in children's relationships with their pets. We call it the Emotional

Energy Checklist. You will use the list to help children discover
and talk about the emotions generated by the death. You might feel
awkward having a printed list in your hand while you talk to your
children about feelings. It might be a good idea to read this list a
few times, put it down, and then talk to your children. When you
and your children near the end of your chat, check the list to see if
there's anything you missed.

Depending on age and maturity, your children will probably
have thoughts and feelings related to some of the categories. We
have created a list that includes a lot of possibilities. The list is only
a guide. Please do not think that children will have memories in all
of the categories. This list is relevant to children of all ages. However, there are a few areas about which you may not feel your children are prepared to talk. As the parent, you get to be the judge of
what is and isn't appropriate for your children. Gruesome details
about death are awkward for many people, children and adults. As
always, your children's interest and questions will dictate how far
they want to go. You already know the range of your children's
language and comprehension skills. Keep in mind that the primary
task is to help children find the emotions that they are experiencing
in response to the death of the pet.

Some children will have elaborate memories from many years
ago. As you talk with them, you also may have memories, not only
about this pet but of pets you had as a child. You may also be
reminded of people who have died or who are no longer in your
life. If you experience emotion, let that be okay with you. Let your
children see that you are human. You are, you know!

We have said that this list is for children of all ages. Your
younger children may need some help in remembering all of the
different things we have listed here. Older children may not want
or require the kind of verbal interactions necessary to help the
younger kids. You can show them or read them the list and let
them think about it. You can suggest that they come back to you a
short time later to tell you about their memories. Some children
may want to use one of their friends or some other person to talk

with about this. Let that be okay with you. At the very least you will be planting seeds for the kind of actions they need to take. You can lead a horse to water, show him the water, and explain how water is made; but in the end your little horse will decide when he's ready to drink. Be careful not to push.

Please do not be in a hurry on behalf of your children and try to move them too quickly. Just as time itself does not heal the emotional wounds caused by the death, we must allow individuals to find their own emotional pace. Our task is mainly to give them healthy choices and the freedom and support necessary to make them.

EMOTIONAL ENERGY CHECKLIST

DEATH OF A PET

As we introduce the idea of an Emotional Energy Checklist, we want to make a very strong statement about its use. The checklist is only a guide. It exists to help you to be able to remind children of the kinds of events that might have produced emotional energy for them. Since this list is a composite, it is unlikely that children will have stored energy and undelivered communications in each and every category. In fact, for some children, there may only be a few of these categories that contain anything they might want and need to talk about.

It is also important to use the list as a way of helping children review their relationship with the pet. In effect, you want to harvest what is there rather than plant things that are not.

You may find it helpful to have this list handy when you talk with your children. There is a section at the end of the list where you can make some notes. For older children, it is acceptable to let them have the list and encourage them to make their own notes.

FROM THE BEGINNING

____Getting permission to have a pet

____Promising to be responsible for care and feeding

____Hopes, dreams, and expectations about the relationship

____Planning, searching, and finding the pet

____The first magical moment or the first conscious memory

____Naming the pet

____Early traumas—housebreaking, crying all night

____Scratching furniture, digging holes

____Early joys—nuzzling, playing

___Sleeping in the child's bed—or not

___Forgetting to feed or clean up

___Table manners—begging and eating "people" food

___The bond of trust—best friends, no secrets

___Ran away—lost and returned

___Indoor or outdoor—possible source of pain about the death

___Fighting with other animals—protecting the yard

___Times at the park

___Friendly to houseguests, or not

___Trips to the vet

LONG-TERM ILLNESS (IF RELEVANT)

___Getting sick—diagnosis, treatment, and medications

___Pain and frustration of watching illness

___The decision to put the pet to "sleep"

___Emotions about the last day

___Accidents and other sudden deaths

___What happened?

___How did child find out?

___Did child see the accident or the result?

___Memorial services

___Disposition of remains

___The days and weeks following the death

After you have this conversation with your children, you might want to jot down some notes. You will be able to use them later to help your children communicate and complete some of the feelings that were part of the details of their relationship with the pet.

NOTES:

_____

_____

_____

_____

_____

_____

_____

_____

_____

_____

_____

_____

_____

_____

CHAPTER 16

What to Do with the Review

CONVERTING EMOTIONAL ENERGY INTO RECOVERY COMPONENTS

We opened part three of the book with the following statement: Completion is the action of discovering and communicating, directly or indirectly, the undelivered emotions which attach to any relationship that changes or ends.

The relationship review is the discovery aspect, which leads to completion. Reviewing a relationship helps children discover the things they wish had ended different, better, or more. The review will also reveal unrealized hopes, dreams, and expectations about the future. Your children will discover things they wish they had said or done and things they wish they hadn't said or done. However, awareness of undelivered and incomplete comments is not enough to make children feel emotionally complete.

After uncovering those areas that are emotionally incomplete, there is one more step before those undelivered emotions can be communicated. They must be converted into one of four emotional categories, which lead to completion.

The categories are simple: apologies, forgiveness, significant emotional statements, and fond memories. You and your children probably already have a working relationship with the concept of apologies. You may have taught them something about forgiveness. The vast majority of emotionally important communications which are not apologies or forgiveness, can be made as significant emotional statements. The final category, fond memories, is an especially helpful one for children. It is a means of saying thank you and showing their appreciation for many positive things they remember about a relationship.

Don't worry, we are going to spend plenty of time explaining the four categories. But first, we will revisit the four-year-old boy and the death of his pet hamster. Do you remember what he said? "Mr. Hamster, you were a good hamster. I'm sorry for the times I didn't clean your cage. I was mad the time you bit me, but that's okay. I wish that you didn't have to get sick and die. I wanted to play with you more. I loved you, and I know that you loved me. Good-bye, Mr. Hamster."

Do you see the apology included in the little boy's comments to the hamster? "Mr. Hamster, you were a good hamster. *I'm sorry for the times I didn't clean your cage.*"

It is easy to guess that this little boy had already been taught something about apologizing. In the moments following the death of his pet, he remembered the incidents about cleaning the cage; without any outside prompting, he apologized. The fact that the little boy apologizes does not mean that he is no longer sad. Nor does it mean that he will not think of that event again. It simply means that he has discovered and communicated an undelivered emotional comment.

We have been socialized to believe things like "What's done is done," "Life goes on," and "No use crying over spilled milk." With those ideas as a foundation, it may be difficult for you to understand that it can be very important and very helpful to communicate something *after* someone has died. It is reasonable to suggest that the boy's apology to the hamster has a value to the boy. To the best of our knowledge, the hamster cannot hear the apology. When you think about this idea, you will automatically realize that the hamster would not have understood the words "I'm sorry" even if the boy had said them to the hamster before it died.

The issue really is a simple question. Who are the apologies for? In the situations where there has been a death, there can be no doubt that the apologies are for the benefit of the person making them.

We often pose this question: If you could have your loved one alive again for just a few moments, what would you say? In addition to "I love you," and "I miss you," most children include apologies, forgiveness, and other important emotional comments among the things they would say if they had a chance.

In the following chapters, we will explain the four important categories: apologies, forgiveness, significant emotional statements, and fond memories.

CHAPTER 17

Recovery Components

APOLOGIES FIRST

We have already mentioned apologies as they relate to the two women friends and to the little boy and his hamster. Now we are going to look at apologies or amends in a little more detail.

Just what is an apology? The definition from *Webster's Ninth New Collegiate Dictionary* that best defines apology for our purposes is: "an admission of error or discourtesy accompanied by an expression of regret." As in, "Mr. Hamster, I'm sorry for the times I didn't clean your cage."

Apologies are necessary whether the error or discourtesy has been an act of commission or an act of omission. Sometimes children have said or done things that have been hurtful to others. Sometimes it's what wasn't said or done that is hurtful. Children need to apologize for things they wish they had said or done differently.

APOLOGIES TO LIVING PEOPLE

Some apologies are best done face-to-face, directly to the person who may have been harmed. When that is possible and advisable, it should be done. Sometimes circumstances do not allow a face-to-face apology, and it must be done on the phone or in a letter.

Sometimes the apology must be indirect. That is, it must not be made directly to the other person. As an example, let's imagine that your child has said some very rude things about Aunt Edith. Perhaps he has said these things to one of his friends. It would not necessarily be appropriate for the child to call Aunt Edith on the phone and tell her, "I told some of my friends that you were really stupid, and I'm sorry." In order to make the apology, the child would have to say the hurtful thing of which Aunt Edith was blissfully unaware. The child might feel a little better and more "complete" for apologizing, but more likely a new "incompleteness" would be created by the rudeness of the comment itself. (In the event that Aunt Edith hears about the rude comment and mentions it to the child, then the child would immediately make a direct apology.)

Completion is the result of the action of issuing an apology as a verbal statement, heard by at least one other person. When it would be inappropriate or harmful to make a direct apology to a living person, it is important to make an indirect apology. Even though making the apology directly would not be wise, it still needs to be said out loud, and it requires that someone else hear it. It is very helpful to write down the apology and then read it to a safe person who will maintain confidentiality.

APOLOGIES TO PEOPLE WHO HAVE DIED

Earlier in this section we talked about the relationship review that occurs naturally following the notification of a death or another type of loss. It is in this review that we discover the things that we need to complete by issuing apologies. The death does not complete what is emotionally unfinished between people—in fact, just the opposite. We are unfinished in exactly those things that we realize never got said or heard or repaired. Sometimes it is a loved one who has died;

sometimes it is a "less than loved one" who has died. In either case there is likely to be a discovery of things that still need to be said.

The fact that someone has died does not cancel our need to complete what is unfinished. Remember the two women who had argued and then parted angry with each other? In the scenario where neither called to apologize and one of them was killed in a car crash, the other was left with an undelivered apology for her part in the argument.

Regardless of spiritual, religious, or philosophical beliefs, when someone dies all undelivered emotional communications need to be made indirectly. That is, the things we need to communicate cannot be said directly. We are not suggesting that people cannot talk to loved ones who have died. What we are saying is that in order for a communication to create a sense of completion, it needs to be heard by another living person.

Here are a few typical examples: Grandpa sends a lovely gift for his grandson's birthday. Mom and Dad remind their son to send Grandpa a thank-you note. The son forgets to write the note, and a short time later Grandpa dies. An element of incompleteness for that child might be the fact that he never thanked Grandpa for the gift. The undelivered communication now becomes an apology that should be made as soon as possible. Because Grandpa has died, the apology will have to be indirect. The son needs to be helped to write and speak: "Grandpa, I really appreciated the birthday gift you sent me. I'm sorry that I never thanked you."

Or: Grandma is sick in the hospital. It is not known that her condition is serious. Mom and Dad encourage their daughter to go with them to visit Grandma. The daughter is very busy and decides not to go. Grandma takes a turn for the worse and dies suddenly. The granddaughter is left with a feeling of regret for not having gone to visit her grandma. Obviously, if she had known what was going to happen, she would have gone to the hospital. In this event, the granddaughter might need to communicate, among other things, "Grandma, I'm so sorry that I did not visit you in the hospital. There are so many things I would have wanted to tell you, like how much I loved you and appreciated you."

In both of the above circumstances, the apologies related to positive events. But incompleteness does not always relate to positive things. It is not uncommon for children to have been mean or sassy or rude. So an apology might sound something like, "Uncle Joe, I'm sorry that I was rude to you," or "Aunt Mary, I'm sorry I laughed at the way you walked."

COMPLETION, NOT MANIPULATION

In simplest terms, apologies are for anything that children did or did not do that might have harmed somebody. The purpose of the apology is to help them be complete with what they have done or not done. When the apology is to a living person, there is sometimes an additional benefit of expanding the communication and relationship with that person. But looking at an apology as a way of getting something from the other person is dangerous. The objective is completion, not manipulation. No one is obliged to accept an apology.

Children can complete their part of an event only by apologizing. The other person's reaction is up to them. As you guide your children, remember to tell them that their apology takes care of what they need to do. This becomes critical when there has been a death, and it becomes obvious that the communication of undelivered emotions can go only one way. The apologies made to a person who has died cannot be answered directly. Completion results from children taking the actions for which they are responsible.

SHOULD PARENTS EVER APOLOGIZE?

Children develop by imitating adults, especially their parents. Therefore, most of what they learn about apologizing comes from adult sources. We think that, in general, parents do not tend to apologize very often to their children. For whatever reasons, perhaps in an effort to maintain law and order in their homes, parents often present themselves as "always right." We do not have any statistics to support this idea, but we believe that parents sometimes perceive, mistakenly, that apologizing would be a sign of weakness. Fortunately, we also have seen a growing shift, especially among younger parents, toward making appropriate apologies to their children.

Indeed, if an aspect of completion is communicated by apologizing to friends and family, then parents can certainly be better teachers by illustrating this and making the first move. This is especially important following losses, when the need to discover and communicate about things we might regret having said or done becomes critical. It will be a lot easier for all concerned if the foundation of the effective use of apologies has already been built.

TIME DOESN'T CREATE COMPLETION, ACTIONS DO

The relationship review is not limited to the hours and days immediately following a loss. In fact, you probably know people who have been telling the same story about a death or divorce for years and years. Essentially, what you are hearing is an ongoing review of a relationship, *without any completion*. The lack of completion is the reason that the story must be repeated over and over.

Children are not much different from adults in this regard: they will also tell a story over and over, trying to find a way to become complete. Try to be alert for those aspects of a repeated story that might indicate that your child needs to make an apology in order to become complete within his or her relationship with someone living or dead.

We have all heard that life goes on, that we must let go and move on, and that we shouldn't dwell on the past. While there is an intellectual truth to those and other clichés, we are not usually taught *how* to move on. If we have never learned how to move on, then teaching that lesson to our children would be difficult. Apologizing is a critical action link on the path to completion. We must help our children discover and deliver, directly or indirectly, the apologies that will allow them to move on.

CHAPTER 18

Recovery Components

FORGIVENESS

Here again is the story we told about the little boy and his hamster. Do you remember the forgiveness included in the little boy's comments to the hamster? "Mr. Hamster, you were a good hamster. . . . *I was mad the time you bit me, but that's okay. . . .*"

Forgiveness is almost always an essential element required to complete the unfinished and incomplete emotions that attach to any relationship. Try as we might, it is almost impossible to keep the slate clean at all times. Because relationships are filled with so many possibilities for misunderstanding and misinterpretation, from time to time we all get our feelings hurt. Children are particularly susceptible to feeling hurt or slighted by others. Children become keenly aware of those collected hurts following a death. As another natural part of the review of the relationship, children discover some things that they believe the other person said or did, or did not say or did not do, that hurt them.

With our little friend and his hamster, we see an almost perfect example of forgiveness in the boy's statement—*I was mad the time you bit me, but that's okay*. The natural wisdom of a four-year-old, who most probably doesn't even know the word or the concept of forgiveness, is demonstrated by the statement, *"but that's okay"*—a

statement of forgiveness that doesn't even need to use the word forgive. *Out of the mouths of babes*, as the saying goes.

It would be wonderful if we could all remain as simple and truthful as the four-year-old. But as we get older and acquire more information, we sometimes lose the elegant simplicity of childhood. Society's concept and experience of forgiveness, both as a practical matter and as a philosophy, are fraught with confusion. In *The Grief Recovery Handbook*, we devoted several pages to explaining the problems associated with many of the ideas passed from generation to generation concerning forgiveness. Because the explanation was so complete, we are going to repeat it here, with modifications, so we can apply it to your ability to help your children.

**Forgiveness is giving up the hope of a
different or better yesterday.**

Forgiveness is one of the least understood concepts in the world. Most people seem to convert the word "forgive" into the word "condone." If we believed the two words to be synonymous, it would be virtually impossible to forgive. Webster's *Ninth New Collegiate Dictionary* definitions illustrate the problem.

> forgive . . . to cease to feel resentment against [an offender].

> condone . . . to treat as if trivial, harmless, or of no importance.

The implication that we might trivialize or easily dismiss a horrible event is clearly unacceptable. However, if we rely on the actual definition of forgive, we would be on the right track.

Before we go on, we want to define the word *resentment*, since it is a major aspect of the definition of forgive. From our trusty old dictionary:

> resentment . . . a feeling of indignant displeasure or *persistent ill will* at something regarded as a wrong, insult, or injury.

We have added the italics to the phrase "persistent ill will" so you

can think about and imagine the impact of holding on to a resentment.

Any resentment etched into the memory of events that occurred in the past will limit and restrict children's ability to participate fully in life. Any reminder of the person or the event about which the resentment exists may stimulate a painful reliving of the unfinished emotions attached to it. Successful recovery requires completion of the pain rather than retention of the resentment.

People's insensitive, unconscious, and sometimes evil actions often hurt children. The continued resentment—and the implicit lack of forgiveness on the part of the child—hurt the *child* rather than the perpetrator. Imagine that the perpetrator has died. Can the child's continued resentment harm the perpetrator? Clearly not! Can it harm the child? Absolutely! As with all recovery components, the objective of forgiveness is to set the child free.

The subject of forgiveness carries with it many beliefs, passed on from generation to generation. Some people have developed such a massive resistance to the word *forgive* that they cannot use it. We recently helped such a woman. She called it the "F" word. We gave her the following phrase: I acknowledge the things that you did or did not do that hurt me, and I am not going to let them hurt me anymore. A variation of that phrase is to say, I acknowledge the things that you did or did not do that hurt me, and I'm not going to let my memory of those incidents hurt me anymore.

We teach children to forgive so they can regain a sense of well-being. Forgiveness has nothing to do with the other person.

FORGIVENESS IS AN ACTION, NOT A FEELING

Many people say, "I can't forgive him; I don't feel it." We agree, because you cannot feel something you have not done. More clearly, you cannot feel forgiveness until you take the actions of forgiveness. A feeling of forgiveness can only result from the action of verbalizing the forgiveness. Action first, feeling follows. If you teach your children the effective use of forgiveness, you have given them a life-affirming skill.

Forgiveness is relinquishing the resentment children hold against another person. They might need to forgive them for something they actually did: I forgive you for ruining my birthday party. On the other hand, they might need to forgive them for something they did not do: I forgive you for not attending my graduation.

There is another expression that people use. "I can forgive, but I can't forget." We have watched people use this awkward combination of ideas to stop themselves from gaining the freedom that is the by-product of forgiveness. Imagine that you had been horribly beaten over a period of many years. It is unlikely that you would ever forget those incidents. But lack of forgiveness of the perpetrator keeps your pain current and alive. Forgiveness does not take away the memory, it takes away the pain.

The implication of "I can forgive, but I can't forget" is that "since I cannot forget, I will not forgive." The real questions are: Who stays in an emotional prison cell? Who continues to resent and shut down their own mind, body, and heart? Whose life is limited by the absence of forgiveness?

Use this new awareness about the value of forgiveness to enhance your children's lives.

We are often asked the following question: In dealing with living people, is it appropriate to forgive someone in person? Our response: Absolutely not! An unsolicited forgiveness will almost always be perceived as an attack. The person being forgiven need never know that forgiveness has occurred. Remember, never suggest that your children forgive anyone directly to their face.

One more note: Many people ask others to forgive them and teach their children to do the same. We think this is an incorrect communication. In fact, it is a manipulation by which you ask the other person to do something that you need to do. And when you ask someone who has died to forgive you, you are asking a dead person to take an action. Children need to take an action, not ask someone else to do it for them. If children ask for forgiveness, they are trying to apologize for something they have said or done. It is

much better for them to make an apology than to ask for forgiveness. That way they can feel more complete.

Let's take everything you just read about forgiveness and look at the little boy's comment to his hamster one more time: "I was mad the time you bit me, but that's okay." In effect, the little boy has issued a perfect statement of forgiveness. The phrase "that's okay," is a statement of perspective. It says that the event happened in the past, and that the little boy is not going to use it to limit himself in dealing with his sadness over the death of his pet.

We are not suggesting that the phrase "that's okay" signifies that bad things are okay. We do not mean to imply that when we encourage children to forgive those who have harmed them, that we are condoning any bad, illegal, or dangerous actions. Far from it. We are saying that they must forgive in order to move on. As long as they stay trapped in the past, they cannot move forward.

At the end of the prior section on apologies, we note that while we are taught the phrases "life goes on," "let go," " move on," and "don't dwell on the past," we are usually not taught *how* to move on.

The topic of forgiveness is actually very simple if we could just stick to that dictionary definition: forgive . . . to cease to feel resentment against [an offender]. But as our children get older, they will inevitably acquire conflicting information from a wide variety of sources concerning this subject. You have to find appropriate ways to communicate about forgiveness to your children, depending on their ages. Remember, the dictionary definition, as well as our ideas, do not condone bad actions by anyone. Nor do they imply that you or your children ever have to see or talk to anyone again. Forgiveness is only one of the tools for removing the pain caused by past events, rather than reliving the pain over and over. The better you impart this idea and skill to your children, the more successful they will be in their lives. We have seen countless lives destroyed by people's inability to complete their relationship to past painful events that happened years or even decades ago.

CHAPTER 19

Recovery Components

SIGNIFICANT EMOTIONAL STATEMENTS

So far we have looked at apologies and forgiveness as two of the major action tools necessary to move toward completion of the pain caused by loss. By now you will probably agree that the absence of apologies and forgiveness can be a recipe for disaster.

The third major communication category that allows children to complete unfinished and unresolved emotions is called significant emotional statements. It is a bit of a ponderous title; perhaps it would be better if we called it really important stuff. Doesn't look as impressive, but it does say it all.

A significant emotional statement is anything of emotional value that is not an apology or a forgiveness. It is any comment that communicates something important that may or may not have been communicated before someone died, or prior to the end of a relationship in the case of divorce or estrangement from living people.

Once more, we return to the four-year-old and his hamster to illustrate significant emotional statements. "*Mr. Hamster, you were a good hamster.* I'm sorry for the times I didn't clean your cage. I was mad the time you bit me, but that's okay. *I wish that you didn't have to get sick and die. I wanted to play with you more. I loved you, and I know that you loved me.* Good-bye, Mr. Hamster."

The little boy's comments actually contain four separate significant emotional statements. First *"Mr. Hamster, you were a good hamster."* You can't get much more significant than that. Then, *"I wish that you didn't have to get sick and die."* While both of those statements has emotional significance for the boy, neither of them requires an apology or a forgiveness. The boy says them because they are emotionally accurate for him. *"I wanted to play with you more."* This is clearly in the category of hopes, dreams, and expectations about the future. And finally, *"I loved you, and I know that you loved me."*

A different little boy or little girl might have said different things, unique to his or her own personality and to the individual nature of the relationship with the hamster. You may remember that little boy had an older sister. She had a much less intense attachment to Mr. Hamster. Her comments following the death were different, because she had a different relationship.

The significant emotional statement category is a convenient catchall for anything that needs to be communicated.

ARE THE SAME THINGS SIGNIFICANT FOR EVERYONE?

What is significant for one child may not be significant for another. Even in a relationship with the same person, thoughts, feelings, and reactions vary widely. We must be careful, especially as parents or guardians, to recognize the uniqueness of every relationship. We must not plant our ideas into the heads and hearts of others. Children are tremendously impressionable, and they will often adopt thoughts and feelings that are not their own.

We have been suggesting throughout this book that you, as the adult, are the emotional leader. We have and will continue to encourage you to go first to demonstrate your response to loss so your children can copy the idea that the expression of feelings is safe. It is probable that you did not have the same kind of relationship with the hamster that your children had. But that does not mean that you would be entirely detached from the event. In fact, there is a very real possibility that the death of the hamster could remind you of parallel losses you had when you were young.

In remembering your own early losses, especially with your pet, or a friend's pet if you didn't have one, you have a natural reservoir on which to draw to help your child discover the thoughts and feelings that have been caused by the death. The Emotional Energy Checklist is designed to give you some very specific guideposts to help you teach your children how to find out what is accurate for them.

SOME SIGNIFICANT COMMENTS REQUIRE FORGIVENESS

The idea of significant emotional statements might seem to be a very open category. But it's not. Sometimes there's a tendency to recite a list of painful statements, as if by saying them, we would be free of their sting. While those comments may be both significant and emotional, they do not create completion. For example, if the child had said only, "I was mad the time you bit me," what would have been communicated is a statement of anger, without any forgiveness attached. As we explained earlier, holding on to resentments by not forgiving is what creates ongoing problems. While it is perfectly acceptable to make a negative comment, it must then be completed with a statement of forgiveness. Let's see if our four-year-old can teach us again: "I was mad the time you bit me, but that's okay."

FOND MEMORIES

In *The Grief Recovery Handbook*, we used only three categories: amends, forgiveness, and significant emotional statements. Here we are adding a fourth category, one that we have been teaching parents for several years because we know it is very helpful for children. Fond memories can be anything that a child remembers with happiness. They can involve sentiments that the child has already said but feels a need or desire to say again. Fond memories can include thank yous and appreciation of positive things. This is an especially important category with pets.

RECAPPING THIS SECTION

This section began with the question Do we know enough yet? Well, we are getting really close. Using the simple but profound example established by the little boy and his hamster, we are beginning to get a picture of the components and language that lead to completion of anything that is emotionally incomplete in the major recovery categories. We are now much more aware of exactly what kinds of things will help a child to move on following a loss.

Just to be a little cautious here, we are not suggesting that every four-year-old would react exactly the same as the one in our story. Even though the story is true, your child may respond entirely differently. And, as we mentioned earlier, the little boy's words do not mean that he will never think, feel, or talk about the hamster again.

Apologies are a part of everyday life. They serve to complete the minor and major disturbances that erupt between people of all ages. We know that one of your tasks as a parent has been to teach your children to recognize when they have been hurtful to others, and to take immediate action to apologize. As you may now realize, the essential goal of apologizing is to complete what has become unfinished by unkind words or deeds. Since you already understand apologies and their underlying importance, you will now be able to help your children apply this idea to those unfinished aspects of their relationship with someone who has died, as well as to other losses.

Forgiveness is a bit more complex. It is undoubtedly more difficult to explain or demonstrate forgiveness to your children, especially the younger ones. That is why we took the trouble to repeat the entire forgiveness section from *The Grief Recovery Handbook*. We strongly recommend that you go back and read that section again. Lack of forgiveness is the largest stumbling block to successful completion of the pain caused by loss. A statement of forgiveness itself is no more difficult than an apology. But the idea which most of us have that forgiveness condones hurtful behavior complicates an otherwise lifesaving action.

Keep in mind that the lack of forgiveness always imprisons the wrong person. We know people who have poisoned their own lives for decades, even though the perpetrator who harmed them died a long time ago. The teaching of effective forgiveness as a daily life tool can be one of the greatest gifts you can give your children. Without forgiveness, children are doomed to a life of almost permanent victimhood, in constant painful memory of things that happened a long time ago.

The all-inclusive category, significant emotional statements, permits the communication of everything else that needs to be said. Positive ideas and things for which we are grateful fit nicely into this area. Statements like, "Grandpa, I always loved when you came to my house and played with my trains. I will never forget seeing you on the floor, with that tiny engineer's hat up on top of your head, while you made the sound of a train whistle. I don't know who was the bigger kid, you or me. Those were very special times for me. Thank you, Grandpa."

That comment was made by a twelve-year-old boy after his grandpa died. As you can see, that communication is a little more sophisticated than the words of our little boy with the hamster. His thoughts and feelings accurately reflect his age and communication skills as well as represent his unique relationship with his grandpa.

"Fond memories" is a very helpful category for younger children. Imagine the Grandpa playing with the trains as remembered by a six-year-old. That boy's statement might simply be "Thanks for playing 'trains' with me, Grandpa."

The freedom to move on is born out of the successful use of all of the actions that lead to completion. As you understand and practice these ideas in your life, you become a better teacher to your children. Major losses do not happen every day. But minor losses and disappointments occur frequently. If you use the principles outlined in the four recovery categories as tools for dealing with loss of all kinds on an ongoing basis, then these tools will become automatic for you and your children if and when a major loss occurs.

CHAPTER 20

Death of a Person

So far, we have focused primarily on the death of a pet. Now we are going to shift our attention to the death of a person. The person might have been a relative, a close friend of the family, or even one of your child's playmates or teachers. In these cases, everything you have learned up until now takes on added importance. The concepts and actions we have been explaining are just as true in regard to the death of a person as they are to the death of a pet. We are going to look at them in the light of some specific real-life stories so you can see how to help your children. Our illustration focuses on the death of a grandparent, partially because of the high probability that this will be the first death of a person that your child will experience.

As we make this shift to dealing with the death of a grandparent or other relative, it is very important to remember that we never compare losses, ever. The death of a pet and the death of a person are not comparable. The fact is that no losses are comparable. If you think about your own life, you might recall hearing about the death of someone you knew, but with whom you did not have a very strong relationship. Most likely the information about that death will not have had dramatic impact on you, precisely because your relationship was minimal. On the other hand, if you have lived with a pet for years, there is a very high probability that the death of that pet will affect you emotionally. Does that mean that you value human life less than you value pet life? No! It simply means that you had a stronger relationship with the animal who died than you did with the person who died.

Grief is about relationships—and we never compare relationships. You may have had a grandma and a grandpa, and you may have loved them both. But we guarantee that you loved them each differently, because each relationship was unique.

You may have had a grandma whom you loved and a grandpa whom you couldn't stand. You may even have had some sense of loving Grandpa just because he was "family" even though you didn't really like him. The keys both to grief and to recovery are an acknowledgment of the uniqueness of each and every relationship. The same holds true for relationships with people, animals, homes, and prized possessions.

What we want to do is help children complete the pain caused by the change or end of the familiar events and activities intrinsic to all relationships. Those events or activities will probably be a mixture of good and bad, and will produce a wide range of memories and emotions.

As we begin to show you how to help children deal with the deaths of people, we want you to be aware that those relationships are more complex because people are more complex. And we must alert you to the fact that, as the parents, your own relationships with the people who have died can affect your ability to help your children.

REVIEWING RELATIONSHIPS WITH
PEOPLE WHO HAVE DIED

The essential ideas about reviewing relationships with people who have died are not different from the review following the death of a pet. We are still looking to help our children discover anything they wish had been different, better, or more. As they remember the events—both positive and negative—in their relationship with the person who died, they will find things they wish they had said or not said. Because relationships between humans tend to be more complex than those we have with animals, there is a higher probability that the categories of apologies and forgiveness will be more important.

The death can also create a painful awareness of the end of any hopes, dreams, and expectations children had about the future, in relationship to the person who died. This can be especially powerful when grandparent, for example, has been at every major event in the children's lives.

THE DEATH OF A GRANDPARENT

You may have been drawn to this book because one of your parents died, or because one of the parents of your spouse died. In either situation, your children have experienced the death of a grandparent. This might be the first human death that has directly affected them.

At this point we want to alert parents that they automatically have very different relationships with their own parents than their children do with those same people. The death of your parent will probably have a large emotional impact on you, as is natural. But it may not have the same effect on your children. As a generality, you have probably been the disciplinarian and rulemaker for your children. The grandparents usually don't have that task, and they often are perceived as the "good guys" who bring or send gifts, and who are sometimes safe people to talk to when kids are struggling with their parents.

It is important to recognize these significant differences. If it is your parent who has died, you will have your own lifetime of emotional energy related to that person. There is some danger of mixing in your relationship with your parents when you try to help your children. If it was your mother-in-law or father-in-law who died, then the relationship may have had less intensity. However, some of you reading this may have had very long-term, wonderful relationships with your spouse's parents. You might relate to them as another set of parents. At any rate, be careful to remember that it is your children's memories of their unique relationship on which you should be focused.

UNIQUENESS IS THE REAL ISSUE

The death of a grandparent is *not* the issue here. The issue is the physical end of a unique relationship. Since all relationships are individual, your children's relationship with that grandparent is the issue. If the relationship was a good one, with a great deal of positive events over a period of time, the death will probably produce a great deal of emotional energy. If the relationship had little contact and few positive or pleasant interactions, the death may not produce a lot of emotion in your children.

If it is your parent who has died, you can expect it will produce much emotion in you, even if you haven't had a lot of contact with your parent over the past several years. It is inevitable that your own childhood memories will be triggered by the death of your parent. Please recognize that your children will not have the same memories and same emotions that you have.

A part of what we are saying is that you must not make the mistake of focusing only on the fact that a grandparent has died, and assuming that your children should have a predictable response. In fact, very young children may not remember the grandparent within a very short time of the death. This is perfectly natural and okay, if you let it be. Years later, through photo albums and other memorabilia, they may become interested again.

The emotions associated with a relationship are the product of both time and intensity. The emotions do not occur just because there exists a blood-relative status. Think about the fact that brothers and sisters within a range of five to six years apart may stay very close over their entire lifetimes—despite the fact that they may have fought constantly when they were younger. When one of them dies, the others are very liable to be strongly affected. If those same close siblings have a much younger brother or sister, they may be much less affected by the death of the younger sibling. That may sound cruel, but it simply points out what we are trying to say: Every relationship is unique, and what the relationship consists of is the emotional key.

People feel the way they feel. Our feelings about other people are always based on the unique events and interactions we have with them. Children are people, too. Their feelings are generated by the special events and interactions, both positive and negative, that occur between themselves and others.

If Grandpa Jones lives in the children's community and the children see him every week, whether the relationship is positive or negative, it will produce a lot of emotional energy. If Grandpa Jones dies, the children will have an emotional response in proportion to the intensity of the relationship.

On the other hand, if your children's grandparent lives thousands of miles away, they may have seen them only once or twice. Yes, they may have talked on the phone many times and received many gifts through the mail, but the contact would have been somewhat limited. If that grandparent dies, your children's emotional response will directly reflect the intensity—or lack thereof—that your children experience within that relationship. Whatever the response, it will be accurate for your children.

Please be careful not to determine how big your children's emotional response should be. Again, if it was your parent who died, you will be dealing with your own emotions. Also, please avoid the trap of implying that, because the person who died was a relative, children must have a very high level of emotional response. Remember that each of your children is different. Each of them had a unique relationship with the person who died. You have your own distinct relationship with the person who has died. Your most important task as a parent is to demonstrate the emotional truth of your own relationship with the person who died; then you can encourage your children to discover the individual truth of their relationship.

"LESS THAN LOVED ONES"

There is a possibility that your relationship with your own parent who died was not a good one. Or, at best, was mixed. At the same time, your children may have had a much better relationship with your parent than you did. You must be careful not to let any of the negative elements of your past affect your children's emotional response to the death.

You also may have a long-running feud with one or more of your in-laws. Again, your children may have a totally different and positive relationship with them. The issue in all circumstances is to make sure that your children are dealing with their unique relationship with the person who has died.

If you can accept the possibility that all parents and children don't have perfect relationships, then you should be able to allow that not all children love all of their relatives. There may be many reasons for this. Those reasons are liable to surface in the relationship review. You must try to remain neutral about your children's relationship with the grandparent who died.

Children have their own reactions to grandparents. Your task is to help your children discover what is unfinished or incomplete for them in their unique relationship with the one who has died. We also must remind you that the more you help yourself in the area of completing what may be emotionally unfinished between you and the person who died, the better you will be able to help your children.

COMPLEX RELATIONSHIPS

As we have mentioned, relationships with people usually are more complex than relationships with animals. We are more likely to get our feelings hurt by a relative than by a cat or dog. We may feel insulted, criticized, or any number of feelings as the result of things said or not said by others. We may also feel loved, adored, cherished, and honored by the words, deeds, and actions of others. Most people, adults and children, have a mixture of feelings, both positive and negative, about all relationships. When someone dies, the emotions attached to the positive and negative events within a relationship can be the source of difficulty and pain for all of us, especially children.

This book is devoted to the idea that children need to discover and truthfully communicate what is emotionally incomplete in those positive and negative events that they remember in relation to the person who died.

Although we use here the example of the death of a grandparent, it is equally acceptable to use the same list when an aunt, uncle, or any other relative has died. It is even okay to use it when a non-family member, a close friend, or a teacher has died.

We have made the categories as broad as we can, and we have tried to make this list as inclusive as possible. You might think of some other life areas that are relevant to your children's experience. Please feel free to add them.

EMOTIONAL ENERGY CHECKLIST

GRANDPARENT, RELATIVE, OR CLOSE ACQUAINTANCE

Again, we remind you that this checklist is only a guide. It exists to help you to be able to remind children of the kinds of events that might have produced emotional energy for them. Since this list is a composite, it is unlikely that children will have stored energy and undelivered communications in each and every category. In fact, for some children, only a few of these categories may contain topics they might want and need to talk about.

It is also important to use the list as a way of helping children review *their* relationship with the person who died. Again, you want to harvest what is there rather than plant things that are not. You may find it helpful to have this list handy when you talk with your children. There is a section at the end of the list where you can make some notes. For older children, it is acceptable to let them have the list and encourage them to make their own notes.

__Meeting or first awareness

__Mom or Dad's parent (or other relationship)

__Special names—Nanny, Poppy, etc.

__They took care of child (baby-sitting) or child stayed at their house

__Punishing or easygoing

__Gifts, or lack of gifts, and/or better gifts to siblings or others

__Trips to their house

__Visits to children's house (kids sometimes lose room to
    grandparents)

__Smells—alcohol, perfume, medication, tobacco

__Grandparent fights with Mom or Dad

__Very safe and easy to be with or talk to

__Scary

__Pinches cheek too hard, teases, embarrasses

__Personal idiosyncrasies—Positive or Negative

LIVES NEARBY

__Frequent visits

__Love the visits or hate the visits

__Don't see much—happy or sad about that

LIVES FAR AWAY

__Don't see much

__Frequent visits

__Lots of phone calls—good or bad

__Few calls—happy or sad about that

__Child observes how grandparents interact with each other

__Child observes how parent interacts with grandparent

__Stay with grandparents when parents on vacation—
    like/don't like

__Wants to live with grandparent when mad at parents

__If local—shows up at school or other events—like/don't like

__If distant—calls re: school events/birthdays/etc.—like/don't like

__Shows up for major events

LONG-TERM ILLNESS

__First awareness of illness—reaction

__Child's observation of parents' reaction

__Diagnosis, treatment, medications, especially relevant if local

__How does parent talk about their feelings regarding the illness?

__Is child allowed or encouraged to talk to grandparent?

__Is child willing to talk?

__Is there anyone else child might talk to?

__Visits and what happens at visits

__When it appears that illness is terminal and how is that
     communicated?

__Is there anyone for the child to talk to about the potential death?

__Do parents pressure the child to visit or call, even if child
     is unwilling?

NEAR THE END

__Circumstances and events that child remembers from
     the last days

__Emotional response (or lack of) to those events

__Was child included in the hospital/hospice care?
     Did child have a choice?

__Anyone safe for child to talk with about what was happening?

__Does child try to take care of parent's emotions?

THE LAST DAY—OR SUDDEN DEATH

__Phone call if far away

__Who told the child and how?

__Emotional impact on child, if any

__Did parents show or express emotion in front of child?

__Child at bedside—home or hospital

__Last conscious interaction—phone or in person

__If in coma—did child speak anyway?

__Anyone safe to talk to about the event?

__Funerals, burials, and other memorial services

__Attend or not? Choice or not?

__Emotional reaction—or not (children often copy adults,
   i.e., be strong)

__Anyone to talk to?

__Days/weeks/months following—dreams, memories,
   regrets, etc.

CHRONICLING EVENTS THAT OCCURRED AFTER THE DEATH

__Holidays, birthdays, any other special days

__Recitals, sporting events, art exhibits

__Graduations, confirmations, bar mitzvah

__Parent's fights and divorce (important, if grandparent was
   safe to talk to)

NOTES:

_____

_____

_____

_____

_____

_____

_____

_____

_____

_____

_____

We have tried to make this emotional energy checklist inclusive enough to open up all of the potential areas in which your children might have stored emotional energy about the relationship. You will probably discover that you, too, have a substantial amount of emotional energy in many of the categories. Please let that be okay with you. It proves that you are human and emotional and honest—all of the things you want your children to be.

RECAPPING PART THREE—IS IT SOUP YET?

We began this section with the question Do we know enough yet? We suggested that even though you have learned a great deal so far, there was still more you needed to know in order to help your children complete the pain caused by loss. Here we are, many pages later, and we hope you are gaining much more clarity about the actions that will help your children. So, is it soup yet? Not quite, but we're almost there.

Talking with your children and helping them remember all the important events and feelings attached to their relationship with pets or people is a powerful experience for children and adults alike. If it were enough simply to remember and talk about pets or people who have died, then we could just stop here and this book would be finished. But talking about a relationship—the good, the bad, and sometimes the ugly aspects of it—does not make us emotionally complete.

It is likely that some time has elapsed since your children experienced the loss that brought you to this book. It may have been some days, weeks, months, or even years since the loss occurred. Not to worry, the actions we will outline in part four will still be relevant and effective.

BEFORE WE MOVE ON, WE HONOR THE READERS

We want to take a moment to thank you and honor you for spending the time and energy to read this book so that you can help your children. It is one thing to want the best for your child; it is something else when doing so means you have to educate or reeducate yourself so that you can guide them. As you can see, this is not a simple, quick-fix idea designed to give short-term platitudes so you can say that you helped your children. Grief is a confusing and often painful collection of emotions. We are glad that you are willing to put out as much energy as you have so that your children can get long-term benefits from what we have learned over the past twenty-five years.

As advocates for children everywhere,
we salute you and thank you.

PART FOUR

Moving from Discovery to Completion

One of the primary purposes of talking about a relationship is to discover the things that we wish we had said or done differently, better, or more. But discovery is not completion, a fact that often seriously trips us up. People often mistakenly believe that awareness or discovery is completion. For example, if you become aware that you have hurt someone's feelings but do not apologize to that person, you remain incomplete. The awareness does not automatically translate into an action that completes what is emotionally unfinished. This becomes especially apparent after someone has died and we recall things that we had never had a chance to say. Remembering things that were not said is a discovery, but it is not completion.

CHAPTER 21

Continuing Litany vs. Freedom

CARRYING THE LITANY IS A HEAVY LOAD

Several times throughout the course of this book, we have referred to those people we all know who have a constant litany of sorrow. In particular, we have mentioned those who repeat compulsively the story about a death, a divorce, or some other loss. We have suggested that the repeated telling of a painful story does not create completion. When you have listened to someone's story for the umpteenth time, you might be tempted to tell them that perhaps they are stuck.

But, think about it: if these people knew how to move on, they would. If they knew how to let go, they wouldn't still be telling the story. They can't turn over a new leaf because they haven't finished with the old one. So the litany goes on, and on, and on.

In our personal lives, we meet the same kinds of people you meet. We hear the same kinds of stories. Sometimes we can get someone alone and ask, "Have you ever thought of forgiving old Whatshisname?" You may be able to imagine the conversations that ensue after we have asked *that* question.

Now, we want you to take any unintentional humor out of what we have just said. Imagine there are children who are telling the same litany over and over. Or worse, imagine these children who, realizing that no one is hearing them or helping them, *stop* talking about the loss, whatever it is, and just bury the emotions inside.

The litanies that children carry inside themselves that limit and restrict their lives are partly the product of the resentments that they cling to only because they do not know how to let them go. Children hold on to resentments because they have observed the people in the world around them holding on to resentments. The incorrect ideas about dealing with feelings come to children from a wide variety of sources. Literature, films, radio, TV, and music are a massive part of the world your children live in. Though often not correctly represented in these mediums, forgiveness is an essential component of completion. Forgiveness allows children to let go, move on, and turn over a new leaf.

Resentments are only one aspect of a litany. Many of the repeated tales you hear children telling reveal the list of things that children wish they had done differently. Often children don't realize that an apology can be issued after someone has died, so they hold on to the pain of the undelivered communication and hurt themselves further. Again, children don't always apologize because they have observed people in the world around them holding on to undelivered apologies. Apologies also allow children to let go, move on, and turn over a new leaf.

A complicating factor of this dilemma is that most relationships contain a mixture of events, positive and negative. If the person who died was occasionally mean to the child, the child may harbor some resentments that could be completed by forgiveness. At the same time, the child may have said or done some things for which he or she needs to apologize. It is not uncommon for a child to be unwilling to apologize for what he or she has done, because of feelings of resentment about the other person's hurtful words or actions. As you can imagine, this can turn into an endless loop, with no conclusion. The result is that the child won't forgive the other person until that person apologizes for what was done. If the other person has died, this apology cannot happen.

The child is stuck, holding on to the pain of the carried resentments, coupled with the burden of the undelivered apologies. Having observed people in the world holding on to this dangerous combination of emotions, a child won't know how to get out of this trap until taught how.

EXAGGERATED MEMORY PICTURES

Grieving people, both child and adult, often create larger-than-life memory pictures in which they either enshrine or bedevil the person who has died or the relationship that has ended.

Once established, exaggerated memories can become a dangerous roadblock to recovery. Since larger-than-life memory pictures are not totally accurate, they make it almost impossible to complete what is emotionally unfinished. For want of a different metaphor, we might say it is impossible to get complete with a saint or with a devil. We must keep in mind that litanies are not limited to bad or negative relationships. It is equally common for people to have a litany about someone they have deified.

Significant emotional statements and fond memories are convenient catchall categories for any important comments that are not apologies or forgiveness. At first glance it might not appear that these categories would produce anything that we might call a litany. However, those categories can sometimes hide another kind of litany. Upon closer inspection, we have known people who have been telling somewhat mythological stories that create a less-than-truthful picture of a relationship with someone, living or dead. In those stories, the litany is of a perfect relationship that did not really exist. As a parent, be alert to overblown positive pictures that might be shielding a layer of unfinished emotions between your children and someone who has died.

Children often do not tell people the important positive things they feel about them because they have observed the people in the world around them holding on to those kinds of sentiments. You have probably heard this comment many times: "I wish I had taken the time to tell him before he died!" It is the communication, indirectly, of those important positive statements that allows a child to let go, move on, and turn over a new leaf.

FREEDOM FEELS BETTER

Freedom is the result of completion. Freedom is the new choice or rediscovered choice available to children when they have discovered and communicated undelivered emotions.

Freedom does not mean the end of sadness, but it can mean the end of pain. Freedom allows fond memories to stay fond and not turn painful. Freedom allows the child to remember loved ones the way he or she knew them in life rather than to be fixated on the images of the loved one in death.

In those instances in which the person who died was a "less than loved one," the child is freed from the haunting memories of the terrible thing that happened. Freedom also can come from completing the pain of the awareness that so many promises had been broken.

For both positive and negative relationships, freedom allows the child to complete his or her relationship to the unrealized hopes, dreams, and expectations about the future. For positive relationships, this includes the sad truth that the person will not be there for the special events yet to come in the child's life. For negative relationships, the child is freed from the often unrealistic hopes that he or she would ever receive an apology that might have mended the damage, and that something of value could have happened.

It is now time to illustrate the very simple idea of converting discoveries into the four major completion categories.

CHAPTER 22

Zeroing In on Completion

"THUMPER"

It is time to show you how to convert the elements of a story into the emotional categories we have been discussing. Our favorite stories come from the people who have learned about grief recovery and so have been able to guide their children when a loss occurs. Our friends Julie and Richard told us how they used the ideas of the Emotional Energy Checklist to help their daughter Jessica. We have highlighted some comments in bold to emphasize certain statements that you will see in Jessica's completion communication to her dog, Thumper.

Jessica was fourteen years old when her ten-year-old Border collie Thumper became ill and died. During the first few days following Thumper's death, Julie and Richard used the Emotional Energy Checklist to help Jessica remember many of the major events and emotions in her life with Thumper. They found the note section at the end of the checklist to be very helpful.

Jessica's parents related the early parts of her relationship with Thumper, as she was too young to remember these things. When she was four years old, her parents would read stories to her. Like most bright four-year-olds, she would memorize the combination of words and pictures on each page, even though she could not really read. If her parents made a mistake, she would quickly correct them. Her favorite book told the story of a little girl and her dog. Before long Jessica began asking her parents if she could have a dog. Her parents responded by saying, "We'll see."

Soon Jessica's requests to get a dog became more frequent and more urgent. Julie and Richard determined that it would probably be very nice for Jessica (and for them) to have a dog. When they announced to Jessica that they had decided that she could have a dog, she was overjoyed.

In the days following the decision, there was great hubbub. Jessica, in rather typical four-year-old fashion, was too excited to eat or sleep and just couldn't wait to get her puppy. Her mom got a book about dogs so they could read and look at pictures to help decide what kind of dog to get. But Jessica didn't want to look at pictures, she just wanted to go the "puppy store" and get her dog. It was as if she knew exactly what kind of dog she wanted and would know the minute she saw him.

Now we begin with Jessica's first real memory. On a Saturday morning, Jessica and her parents piled into the car and went to a pet adoption facility, which had many dogs and cats waiting to be taken home by someone who would love them. She remembers walking down a long aisle of cages and seeing many dogs looking out at her, almost begging her to take them. But she just glanced at each one and moved on. All of a sudden she stopped dead in her tracks. In the cage in front of her a little fluff ball was asleep, right in the middle of the noise and confusion all around it. Jessica said, **"Here's my dog."**

On the way home in the car, Jessica sat with a cardboard box in her lap. In the box was her precious cargo, her new best friend. As they drove, her parents asked her if she had thought about a name yet. Just about then, the puppy woke up. As he woke up, his little tail starting wagging, and banging against the cardboard box. **Thump, thump, thump. Jessica blurted out, "Thumper," and the name stuck.**

The next several weeks are a blur for Jessica. Among her memories are those first few nights when **Thumper would cry,** and she would pet him and talk to him and tell him that everything was going to be okay. She also remembers trying to housebreak him, and how excited he would get and forget and "piddle" on the carpet. Jessica's mom would get upset when her carpet got dirty, and **Jessica remembers being scared that Thumper would have to go back to the adoption store.**

Jessica and Thumper became best friends. She started dressing him in some of her doll's clothes. She has some pictures of him dressed in some pretty absurd outfits. While remembering the times she had dressed Thumper, she started crying. **She remembered a couple of times when she had made him yelp trying to force his little dog legs into doll's clothes.**

Not very long after Thumper came to live with her, Jessica started sneaking him into her bed with her, after her parents had already kissed her good night and turned off the lights. It wasn't long before her parents discovered what she was up to. Jessica and her parents had a big meeting. It was very scary for Jessica. She thought she would be in big trouble for sneaking Thumper into her bed. But her parents were pretty understanding, and after a mild scolding for being sneaky, and after promising to clean up if there were any messes, **she was allowed to have Thumper sleep with her every night.**

When Thumper was about six months old and very frisky, he bolted out of the door one day and ran across the street. Jessica and her mom ran after him and luckily caught him before anything tragic could happen. **It was a terrifying memory for Jessica.** Thumper had never been off a leash and did not know how to deal with streets and cars.

The events that Jessica remembers correspond to many of the general categories on the Emotional Energy Checklist. In fact, of the first eleven items listed in the checklist, six of them are part of the story you have read so far. You will notice the asterisks in two of the categories. Those are memories that are very vivid for her parents, but Jessica doesn't really remember them. We point this out here so you can be reminded that the focus here is your child's memory of his or her pet. Let's review:

* Getting permission to have a pet
__ Promising to be responsible for care and feeding
__ Hopes, dreams, and expectations
* Planning, searching, and finding the pet
✓ The first magical moment or the first conscious memory
✓ Naming the pet
✓ Early traumas—housebreaking, crying all night
__ Scratching furniture, digging holes
✓ Early joys—nuzzling, playing
✓ Sleeping in the child's bed—or not
✓ Ran away—lost and returned

Let's return to Jessica's memories and see how the rest of the story relates to the items on the checklist.

By the time Jessica was six years old, feeding Thumper had become one of her chores. Generally she was pretty good at remembering to feed him and to keep his food and water dishes clean. Every once in a while, she would forget, and her mom or dad would have to remind her. One painful memory related to a time that she forgot to feed Thumper. Later that night he tore into the garbage and made quite a mess. **The next morning Jessica's mom was pretty upset with both Jessica and Thumper. Jessica felt terrible.** She didn't like the idea of Thumper getting in trouble because she hadn't fed him.

On a happier note, Jessica remembered hundreds of times when she and Thumper would be in her room playing, and she would talk to him. She shared with him all of the thoughts and feelings she was having at every stage of her life. **He always seemed to understand.** Sometimes, as she talked to him, he would cock his head to one side, as if trying to hear everything she was saying. He never seemed to mind how much she talked to him and never seemed to disagree with her.

Thumper had adopted one of Jessica's rag dolls when he was a puppy. For years, he would drag his baby around with him everywhere he went. Eventually the doll was just a bunch of tattered threads. One of Jessica's fondest memories was every time the family had company, **Thumper would bring his rag doll baby to each and every visitor and plunk it down in front of them.** Each person in turn had to admire his baby.

When Thumper was still a puppy, he had his first visit to the veterinarian. Although Jessica didn't remember this visit very well because she had been so young, her mom did remember, and told Jessica that Thumper had been very calm. **Thumper established a relationship with the vet**, and even though he would yelp sometimes during an exam, he always seemed to trust the vet. Jessica remembers visits to the vets as she got older.

Let's stop again and look at Jessica's story as it relates to the Emotional Energy Checklist. Notice that there are check marks in four of the categories.

✓ Forgetting to feed or clean up

__ Table manners—begging and eating "people" food

✓ The bond of trust—best friends—no secrets

__ Indoor—outdoor, possible source of pain about the death

__ Fighting with other animals—protecting the yard

__ Times at the park

✓ Friendly to houseguests—or not

✓ Trips to the vet

Jessica's relationship with Thumper was filled with many of the wonderful interactions that are so special between children and animals. But, as her story continues, she remembers a series of events that led up to Thumper's death. Shortly after Thumper's tenth birthday, Jessica noticed something wrong. Thumper was lying on the floor in Jessica's room. **She called to him, but he didn't jump up and run over to her as he usually did.** Instead, he kind of looked up at her, as if to say, "I heard you, but I can't get up right now." She called him again, and this time, slowly, he got up and came over to her. She patted his head and forgot about it until the next day. On the following day, it took three times before Thumper got up and came to her. **Jessica became concerned. She told her mom and dad, and the next day they took Thumper to the veterinarian.**

Jessica looks back on that trip to the vet as one of the scariest days in her life. Even as the doctor was examining Thumper, she sensed that something was terribly wrong. When the doctor asked Jessica and her parents to join him in his office, she couldn't breathe. The next several minutes were a total nightmare. Jessica felt as if her heart were frozen. The doctor told them that he thought Thumper was very ill but that he had to wait for the results of blood tests to be sure. Jessica remembers sobbing into her mom's shoulder and hearing the doctor's voice sounding far away.

Jessica did not eat dinner that night. She also could not sleep that night. She spent the whole night lying next to Thumper, staring at him, stroking him, and saying "I love you," over and over.

A few days later the doctor called and confirmed everyone's worst fears. Thumper had cancer. Jessica and her folks made another trip to the vet for the meeting that would tell them the choices they were going to have to make. The doctor explained the treatment options and the chances of Thumper getting better. Jessica remembers being almost numb as she sat there trying to know what was the right thing to do. She did not want Thumper to suffer. But as long as there was even a chance that the treatment could be successful, she believed that they had to try it. Her parents agreed.

The next several weeks were a whirlwind of medication, tests, signs of improvement, followed by hope, and then bad signs and different medications and more tests. Jessica's emotions went up and down like a roller coaster. Hope, then terror, then hope again.

One morning, when Jessica woke up and looked at Thumper,

she knew that something had changed. His eyes seemed different. When he looked at her, it was almost as if he didn't recognize her. Heartbroken, she lay down on the floor next to Thumper, put her hand gently on one of his paws, looked at him, and cried and cried.

Jessica forced herself to get up, and she walked down the hallway in a daze. She knocked on her parents' bedroom door and asked them to come to her room. Her parents saw the same thing that Jessica had seen, that Thumper could not fight anymore. The three of them stood in the doorway of Jessica's room, looking down at Thumper, who no longer had the strength to look back up at them. His eyes glazed, he just seemed to be looking off into nowhere. As they held on to each other, they talked about what was probably going to happen next. They agreed that they would not want to do anything to extend any pain or discomfort Thumper might be experiencing.

That last car ride to the vet was very quiet. Jessica and her mom and dad were each wrapped in their own personal memories of the past ten years with Thumper. Finally, the car pulled up in the vet's parking lot. Dad turned off the engine. Nobody moved for what seemed like an eternity to Jessica. None of them wanted to do what had to be done.

The doctor made a last examination just to be sure. He explained to Jessica and her parents that they had to choose between having the doctor give Thumper a shot that would end his suffering, or having the doctor continue to administer pain-killing medications. A third choice was to stop the medications and let nature take whatever time it might until he died.

The doctor excused himself to give them time to decide. They agreed to have the doctor administer the shot. They called the doctor back into the room and told him what they had decided. The doctor said that he thought that was the most loving decision they could make. He left the room so they could spend a little private time with Thumper.

Jessica asked her parents to allow her a few minutes alone with Thumper. As she recalls, that was the most emotional experience she had had in her life up to that point. Many of the memories contained in this story came back to her during the time she spent in that little room with Thumper. **She cried, she laughed, and she told him how much he meant to her.** Every once in a while, for just a second, he would manage to look up at her as if to say, "I know, me, too." Finally, she leaned down and kissed him between his eyes and then whispered, "**I love you. Good-bye, Thumper.**"

When the doctor came back, he explained that each of them would have to decide if they wanted to be in the room with Thumper when he administered the shot. He told them that some people had chosen not to be in the room and later regretted that decision. He told them that if they got uncomfortable, they could leave the room. They all decided to stay.

The last scene at the vet's was strangely peaceful for Jessica. She clung to her parents, her heart heavy and her face wet with tears. Thumper did not seem to be in pain, and then, gently, he was gone. Jessica leaned down and kissed him one last time and said, "**I love you, I will miss you, my friend. Good-bye Thumper.**"

The rest of that day is mostly lost in Jessica's memory. She recalls being in her room, lying on her bed, staring blankly at the ceiling and then dozing off for a while. She would wake up with a start and, remembering that Thumper was gone, sob quietly as the reality sank in. She has some idea that her family had dinner that night, but she doesn't think that she was able to eat. Later, Jessica spent several hours going through her collection of pictures. She pulled out all of her favorite pictures of Thumper.

Let's look at the other categories on the Emotional Energy Checklist to see which areas are part of Jessica's story.

LONG-TERM ILLNESS

✓Getting sick—diagnosis, treatment, and medications

✓Pain and frustration of watching illness

✓The decision to put the pet to "sleep"

✓The last day

As you can see, there are check marks in each of the four categories. And you will recall that Jessica's story contains vivid details within each category. Depending on the age of your child at the onset of a pet's illness, there is a high probability that there will be a great deal of energy attached to the circumstances that lead up to and include the death of the pet.

You have seen the results of using the Emotional Energy Checklist to help Jessica review her relationship with Thumper. Next, we are going to show you how those discoveries were turned into completion.

CHAPTER 23

Delivering, Completing, and Saying Good-bye

The four-year-old in our earlier story was able to verbalize a few sentences and complete the things he needed to say to his hamster in the moments after the hamster died. The little boy's mom was familiar with the principles of grief recovery. She knew that it would be important for the boy to repeat the things he had said to the hamster alone in his room. She knew that those kinds of thoughts and feelings always need to be spoken out loud and heard by other living people to be completed communications. At the memorial service in the backyard, Mom helped him repeat those things he had said in his room.

Because he was able to express exactly what was emotionally important to him, there was no need to consult an Emotional Energy Checklist. Because he was young and had a relatively short relationship with his pet, there was no need to write down the things he remembered.

Jessica, at age fourteen, and with a ten-year relationship with Thumper, was helped by her parents' awareness of the relationship review and the Emotional Energy Checklist. Since her story was much longer and more complex, it was helpful for her to write down the things she needed to say to Thumper and to make sure that her list contained the recovery components that would help her be complete.

The best way to consolidate the collection of emotions she had discovered in the relationship review was to put it in the form of a Grief Recovery Completion Letter. The Grief Recovery Completion Letter is not an ordinary letter. It is not a newsletter or a journal or diary entry. It is a very specific kind of letter that communicates the apologies, forgiveness, significant emotions, and fond memories that are contained in a child's relationship with an animal or a person who has died.

There are several purposes for the completion letter. One is to help a child say all the things he or she needs and wants to say so that those thoughts and feelings do not get trapped inside. We talked earlier about children and adults who keep telling the same story over and over. It is usually because they did not have an effective way to complete the emotional discoveries they made naturally following the loss.

The letter also allows the child to say good-bye to the physical relationship that no longer exists. Death ends the physical relationship, while the emotional and spiritual relationships continue. Completing the emotional aspects of the relationship allows the child to say good-bye to the physical relationship.

LEADING UP TO JESSICA'S LETTER

With the help of her parents, Jessica had reviewed her relationship with Thumper and discovered some things that she needed to communicate. Some of the things she needed to say were things she had said many times, yet she needed to say them again. Some were things she had to say for the first time.

Throughout this book, we have mentioned that there are always things we wish had been different, better, or more. The Emotional Energy Checklist helped Jessica find exactly those things that needed to be said, either for the first time or again.

As we have said before, relationships with pets tend to be more unconditional and less complicated than relationships with people;

therefore, there is much less need to apologize to and forgive our pets. There will most likely be a lot of statements of appreciation and of thanks for all the special time and unique communication that occurs between humans and animals. However, we will be on the lookout for those few times when apologies and forgiveness are relevant.

To remind you, here are the four basic categories, which will help children communicate undelivered emotional thoughts and feelings.

> [A] = Apologies—"I'm sorry . . ."
>
> [F] = Forgiveness—"I forgive you for . . ." Or something bad happened, but "that's okay."
>
> [SES] = Significant emotional statements that are not apologies or forgiveness but that definitely need to be said. They could also be called really important stuff.
>
> [FM] = Fond memories—"Thank you for . . ." or "I appreciate . . ." This is an especially important category with pets.

Here is the list of categories that Jessica and her parents talked about, in which important things had happened during the years of her relationship with Thumper.

✓ The first magical moment or the first conscious memory

✓ Naming the pet

✓ Early traumas—housebreaking, crying all night

✓ Early joys—nuzzling, playing

✓ Sleeping in the child's bed—or not

✓ Ran away—lost and returned

✓ Forgetting to feed or clean up

✓ The bond of trust—best friends—no secrets

✓ Friendly to houseguests—or not

✓ Trips to the vets

LONG-TERM ILLNESS

✓ Getting sick—diagnosis, treatment, and medications

✓ Pain and frustration of watching illness

✓ The decision to put the pet to "sleep"

✓ The last day

Jessica's parents suggested that she take the list of events and write a letter to Thumper. Not just an ordinary letter, but a very special letter. A letter in which she could thank him and apologize to him and forgive him. She went to her room with the list, a pad of writing paper, and a box of tissues.

Jessica wrote and cried, and wrote some more and cried some more, and finally came to the end of her letter. With her permission, we have reprinted it here. At the end of each paragraph we have added abbreviations to help you see which category or categories of completion are in that statement. A for apology, F for forgiveness, SES for significant emotional statements (really important stuff), and FM for fond memories and thank-yous.

You will notice that some sentences of Jessica's letter begin with the name Thumper. Some people find benefit in starting some of their communications with the name of the person (or animal).

JESSICA'S COMPLETION LETTER TO THUMPER

Dear Thumper,

I have been remembering our time together and I have found some things I want to say.

One of my happiest memories is the day I first saw you at the puppy store, curled up in a ball. Although I was really little, I remember that day, and riding home with you in a box in the car, and your little tail thumping against the side of the box. [FM]

I remember when you were brand new, I guess you were scared, and you would cry during the night. I remember being afraid that my parents would make you go back to the store. I'm so glad that you stayed with me. [SES]

I'm sorry that I sometimes hurt you, trying to squeeze you into my doll's clothing. I'm sorry for the times that I accidentally stepped on your paws. And I'm sorry for the times I got mad and yelled at you. [A]

Thumper, one of the most wonderful things in my life has been that you always slept with me. I felt so comfortable and safe with you beside me. Thank you for being there. I'm so glad that my parents let us do that. [SES + FM]

I remember when you were still little and very frisky, that you ran out the door and out into the street. I screamed, and my mom and I ran after you. I was so afraid that you were going to get hurt. It was the most scared I had ever been. I forgive you for running away and scaring me. And I am so glad that you didn't get hurt. [F + SES]

One of the worst things was the time I forgot to feed you and later that night you tore into the garbage and made such a mess. I'm sorry that I didn't feed you, and you got into trouble with mom, just because I didn't take care of you. I'm sorry that mom yelled at you, it was my fault. [A]

Thumper, mostly I remember the hundreds of times I would just lie on my bed and talk to you. I told you everything, no secrets. It made me feel so good to be able to say what was going on with me. It helped me figure a lot of things out. Sometimes you would turn your head to one side as if you were trying to listen better, or as if you weren't sure what I meant. You always listened to everything I had to say, and I thank you for that. [FM + SES]

Thumper, you really were very funny. Your baby, my rag doll that you adopted when you were little, created more laughter than anything I ever saw. The way you would plop your doll down in front of everybody, and then look up at them and then down at the doll, as if they were supposed to notice what a beautiful baby you had, was hysterical. I'll never forget that. [FM]

Since you died, I have remembered a lot of things. I remember how much you loved all my friends. I remember how excited you would get every day when I came home from school, as if just the fact that I had showed up was the most exciting thing in the whole world. Thank you for being you and helping me feel really great about being me. [FM]

I remember that day when I called to you and you looked up at me but didn't come right away. I had never seen you do that before. When it happened again the next day I got really scared. [SES]

All the trips to the vets and the terrible day when we found out you had a cancer were so hard. It was hard, because I did not ever want you to be in pain or to have to go through all of those medical things. I wanted to protect you. I'm sorry that you had to suffer. [SES + A]

Thumper, maybe because I was so little when you came to live with me, I had never thought of life without you. Near the end, when we knew that your body couldn't get better, I started to think about what it would be like if you weren't here. I didn't want to think about it, but the ideas would just jump into my head. I would cry because for almost as long as I can remember, you were always there. It has been very difficult, waking up and then remembering that you are not here. I miss you. [SES]

I am old enough to understand that all living things die, and even that I am going to die someday. But, knowing that does not make my heart feel better. It does feel good to tell you all of these things. It feels good to tell you that I remember the thousands of times we had together. It feels so warm and fuzzy to remember you lying next to me at night, or listening while I told you about the first boy that I had a crush on when I was in the sixth grade. Thanks for listening, thanks for loving me. [SES + FM]

I love you. I miss you. I will never forget you

Good-bye, Thumper

Note: Some children might choose to begin each thought with the name, as Jessica does in several of the paragraphs here with "Thumper." This can act as an emotional stimulus in both the writing and reading of the letter. Other children won't do this. Either way is fine.

After Jessica finished writing her letter, she showed it to her parents. Then her mom asked her if she would be willing to read the letter out loud while her parents listened. She was willing. Her mom suggested that Jessica close her eyes for a moment to get a picture of Thumper in her mind's eye before reading.

Jessica closed her eyes, and the image of Thumper and her rag doll jumped into her head. Jessica smiled and started to read. As she read the letter, sometimes she smiled and sometimes she cried. Her mom and dad sat a few feet away and listened. They both had tears in their eyes.

When Jessica came to the end of the letter, she paused. After she said, "I love you. I miss you. I will never forget you," she stopped. Tears trickled down her face, and then she said, "Good-bye, Thumper." As she said those final words, more tears came. Her mom got up quickly, gently pulled Jessica up from the couch, and embraced her in a hug. Jessica wept and wept. After a while, her mom let go and her dad continued the hug.

When Jessica stopped crying, her mom asked her to close her eyes for a second to see if she could still get a picture of Thumper. She closed her eyes and almost immediately began smiling. She told her mom and dad that she could still see Thumper. Her mom did that to show Jessica that she didn't "lose" Thumper by saying good-bye.

Having done the review and written and read the letter did not mean that Jessica was not sad anymore. Nor did it mean that she did not miss Thumper. What it meant was that she had completed the things that she needed to say so that she could learn to adapt to the new circumstances of her life without Thumper physically present.

Her mom told her some of the other things she had learned when using *The Grief Recovery Handbook* to deal with one of her own losses. For example, she taught Jessica that each time she remembered one of those same things she had written in the letter, she could just close her eyes and say them again. And each time she did that, she could say, "I love you. I miss you. I will never forget you. Good-bye, Thumper." Mom explained that it's perfectly okay to repeat those things, and that it is very important to finish each communication with "good-bye."

ENTIRELY DIFFERENT BUT EXACTLY THE SAME

We never compare losses, and we never compare completion communications. But just this once, let's break that rule and compare Jessica's letter to Thumper with the four-year-old boy's statements to Mr. Hamster.

Here are the questions:

Does each communication reflect a unique and individual relationship?

Does each communication address the emotional truth for each child?

Does each communication complete unfinished or undelivered emotions?

Does each communication end with good-bye?

You already know that yes is the answer to all of those questions. The little boy said everything that he needed to say in just a few sentences, while Jessica's communication took a few pages. But each communication represents a total truth for each child.

Between a four-year-old and a fourteen-year-old, there is a world of difference; yet the essential truth is represented in a very similar way. Every child is unique, and every relationship is unique. A fourteen-year-old could have written a four-line letter and a four-year-old could have gone on for much longer. But the odds are that what we have shown you represents the norm. A nine-year-old's communication might fall somewhere in between these two. For you, the most important thing to remember is that your task is to help your children find what is accurate for them.

The next chapter is going to illustrate the relationship review and completion letter written by a teenage girl to her grandmother who has died.

CHAPTER 24

Very Close to NaNa

This is the story of a girl named Amanda and her grandma, whom she called NaNa. Amanda and her mom and dad told the story to us after NaNa died. A couple of years prior to NaNa's death, Amanda's mom and dad had participated in a Grief Recovery Personal Workshop. Because they had an awareness of what is normal and natural, they were able to help their daughter when her grandma died. Again, we have highlighted some phrases in bold to illustrate the elements of the relationship review that led to Amanda's completion letter to NaNa.

Amanda was born in Philadelphia. Rhonda and Jack, her mom and dad, were thrilled and excited in the days leading up to her birth. She was to be their first child. **Rhonda's mom** lived in Florida, but wild horses couldn't have kept her away from the arrival of her first grandchild. She flew up to Philly to be a part of the blessed event and to give her daughter a helping hand during the months that followed.

Amanda does not remember the first moment she saw her grandma, **but she has a sense that Grandma was always there**. Grandma spent a lot of time with baby Amanda. Over the years, that special bond that was established when Amanda was a newborn seemed to grow and grow. Grandma did not like traveling, but each time she got back to her home in Florida she began planning her next trip to Philadelphia. Grandma was constantly saying, "I'm going to Philadelphia to see my little Amanda." Grandma's friends began teasing her. They said the city should be called "Amandelphia."

As soon as Amanda began to make reasonably intelligible sounds, her mom would put her on the phone to gurgle to Grandma. Amanda did seem to recognize Grandma's voice through the phone. It was on one of those phone calls that Amanda said something that **sounded like NaNa.** It stuck, and from that point forward, Grandma became NaNa, and Grandpa became PaPa.

NaNa spent almost all her free time in the baby clothes section of all the stores within a wide range of her house. NaNa may have singlehandedly been responsible for the success of UPS, as **she sent gift packages from Florida to Philadelphia nearly every day.**

By the time Amanda was four years old, she and Grandma were incredibly close. As much as Grandma couldn't wait to get back to visit, Amanda was just as eager to see her NaNa. It is right around this time that Amanda's conscious memory kicks in. Even though she has a sense that her NaNa has always been there (which is true), and even though her parents have told stories over and over about Amanda and her grandma, her real memories start when she was four. In fact, she has some memories from her fourth birthday party. Of course, NaNa had flown up for the event. Amanda remembers introducing NaNa to some of the other children at the party. She remembers that one of the other children said, "What's a NaNa?" Amanda's mom told the kids about the time on the phone when Amanda had first said NaNa. Some of the other kids told how they had come to say "granny" or "nanny" or "gran."

NaNa's frequent visits were very much anticipated and appreciated by Rhonda and Jack. These visits gave them the freedom to catch up on the movies and their social life outside of the realm of little children. For Amanda, those times were special. NaNa was her very best friend. **NaNa never punished or scolded.** NaNa always seemed so comfortable and safe.

When Amanda was very little, the family visited NaNa and

PaPa in Florida. Amanda does not remember that trip. She was only two years old. **She does remember a trip when she was six.** She remembers being very excited because after they got to NaNa's house, her parents were going off on a cruise and she was going to have a whole week with NaNa and PaPa. While most of this story is about Amanda's relationship with NaNa, she also has a wonderful relationship with PaPa.

As she looks back on that trip, Amanda believes that it marks a change in her relationship with NaNa. It was during that week, that **Amanda remembers having really long talks with NaNa.** Not just playing or shopping but really talking. Amanda remembers asking NaNa about NaNa's childhood. NaNa's stories fascinated Amanda. Because of the kind of talks they had, Amanda started to feel bigger, more like a little girl than a child or an infant.

Another memory that sticks out for Amanda is how NaNa and PaPa were with each other. Most of the other times, when Amanda was younger, NaNa would be in Philadelphia, while PaPa stayed in Florida. Sometimes PaPa was there at holidays, but with the excitement and other people around, she never really saw NaNa and PaPa together. On this trip she saw how they were with each other. **The word she uses most to describe how they were to each other is "sweet."** They were considerate and loving, and Amanda felt very comfortable being around them. She even remembers getting a sense of the ways in which her mom, Rhonda, was like NaNa, and other ways in which her mom was like PaPa. It was this awareness that prompted Amanda to start wondering whom she was like, and to start asking NaNa all kinds of questions. Of course NaNa always said that Amanda was special and unique.

The next eight years were more of the same. NaNa came to Philadelphia as often as she could. At least once a year, **Amanda would get to go to Florida to have some private time with NaNa and PaPa.** NaNa managed to **show up at many of Amanda's school events, and if she couldn't make the trip, she never failed to call.**

Most of the time, Amanda felt wonderful about NaNa. They were so similar in so many ways that sometimes it seemed as if they were using only one head and one heart. Except when it came to neatness. NaNa was a neat freak and Amanda was a free spirit. Clothes, dolls, and toys stayed where they landed. Because Rhonda had been raised under NaNa's rigid rules of neatness, she didn't seem to mind Amanda's casual relationship toward keeping things tidy.

NaNa and PaPa lived in a small condominium in southern Florida. They had converted a den into a permanent room for Amanda. NaNa had gone to great lengths to decorate it for her precious granddaughter. It was on that trip to Florida when Amanda was six that the issue of neatness first came to a head. In looking back, Amanda thinks that when she was much younger, NaNa was willing to let Amanda's habit slide because she was just a baby. But, once she hit age six and stayed at NaNa's house, the rules seemed to change. **It was the first time Amanda had seen a side of her NaNa that she did not understand or love.**

One morning during that trip NaNa came into Amanda's room and saw clothes strewn all over the floor. Amanda had been having a hard time deciding what to wear. NaNa hit the roof, or as much of the roof as Amanda had ever seen. Because Amanda wasn't accustomed to her NaNa having a temper, she was taken aback when NaNa's voice went up and her attitude was not very loving and accepting. Maybe that's why this particular event stood out in Amanda's memory after NaNa died.

In a conversation with her mom, back at home in Philadelphia, Amanda discovered the bond she and her mom shared and learned why they both were not very neat. **Her mom told her that it was not very likely that NaNa was ever going to change,** and that Amanda would have to figure out a way to deal with the situation, especially when she was in Florida. From that point onward, Amanda did her best to be a little bit neater at NaNa's house. But, no matter how she tried, she never sensed that she was neat enough to suit NaNa. It was the one area that she could never talk about with NaNa. For all of NaNa's flexibility in other areas, she was rigid when it came to neatness.

It was not until NaNa's funeral, when Amanda had a long talk with her great aunt Sylvia, NaNa's sister, that Amanda found out why NaNa was so stuck on neatness. According to Aunt Sylvia, she and her sister had been punished when their rooms did not meet their parents' rigid standards. The harshness of that treatment had a lifelong impact on NaNa.

Amanda's thirteenth birthday was a big event. NaNa made the trip from Florida to join in the celebration. Amanda's closest friends already knew and loved NaNa. If you think NaNa had some trouble with one messy granddaughter, just imagine her reaction to a gaggle of thirteen-year-old cyclones. By then, Amanda could at least have a little fun with NaNa, especially at her own house where NaNa was not the boss.

The day after the birthday party Amanda and NaNa went for a very long walk and talked. NaNa wanted to know more about Amanda's visions for the future. Amanda told her that she had become more and more interested in art. The art classes she had been taking were starting to pay off. She felt more confident that she could translate the images from her mind's eye onto paper. She was going through a period when she made charcoal sketches of anything and everything. **NaNa loved listening to Amanda talk about her drawing.** When they got back to the house, Amanda showed NaNa her collection of drawings.

When NaNa finished looking at the drawings, Amanda asked her to close her eyes and hold out her hand. NaNa did as instructed. Amanda put something between NaNa's fingers, then told her she could open her eyes. When NaNa opened her eyes, there was a piece of art paper in her hand. **On the paper was a charcoal sketch of NaNa.** NaNa sat and stared at the drawing, tears streaming down her face. Amanda leaned down and hugged NaNa.

As fate would have it, that was the last interaction that Amanda and NaNa would have.

Later that day, Amanda went to soccer practice. NaNa and Rhonda had tea in the kitchen and then NaNa decided to take a nap. A few minutes later, Rhonda heard a horrible gagging sound coming from NaNa's room. Rhonda raced in and found her mother in the throes of a heart attack.

Rhonda screamed and ran to her mom, then grabbed the phone next to the bed and dialed 911. The paramedics arrived very quickly, but it was too late. NaNa had died.

In the bedlam that followed, Rhonda called Jack at his office. Following that frantic phone call, Jack hung up, got into his car, and sped to the soccer field to find Amanda. Jack shudders as he recalls that trip. He adored his mother-in-law, so his heart was broken for his own relationship with her. He was very concerned for his wife, not only because her mother had just died, but because she had just experienced such a horrifying situation. And, most of all, he could not imagine what this news was going to do to Amanda. He did not have any idea how to tell his daughter what had happened.

Jack pulled his car into the parking lot at the soccer field. He ran toward the field. **Amanda saw him running and knew something had to be wrong.** For a split second her mind went to her mom, and then immediately she knew it had to be NaNa. As she remembers those moments, she remembers feeling as if she had been punched in the stomach. She moved toward her dad in what felt like slow motion. She knew it had to be horrible news. There was no other reason for her dad to be here and for him to be running.

When she got a few yards from her dad, she said, "NaNa?" He could only nod. She said, "Dead?" Again he could only nod. She fell into his arms and sobbed. By then the game had come to a halt. Amanda's teammates and coaches stood a few feet away. Many of them had known NaNa.

Amanda and Jack walked to the car holding each other up. When they got into the car, Jack had to take a few minutes to dry his eyes so that he could drive safely. As Amanda recalls, she couldn't feel her body. **She was numb.** As they drove home, Amanda asked her dad to tell her what had happened. Jack could only repeat what Rhonda had told him on the phone, that NaNa had had a massive heart attack and that the paramedics had not been able to save her.

Trying to deal with the crushing reality of her mother's death, and at the same time handle the details with the paramedics and others who were swarming through her house, was overwhelming for Rhonda. She called The Grief Recovery Institute. John and Russell helped her to begin to talk about her emotions surrounding her mom's death so that she could go back and finish up with the arrangements that had to be made. John and Russell reminded her about the review that was already happening naturally, and they gave her some pointers about the Emotional Energy Checklist to help Amanda.

After Amanda and Jack arrived at the house, there was a great flurry of activity. Many more phone calls had to be made to take care of all the details attendant to NaNa's sudden death. Rhonda had waited for Jack and Amanda to arrive so they would be with her when she called PaPa in Florida to tell him the tragic news. Then there were many calls to other relatives and friends. The three of them remember feeling that afternoon and evening as if they'd been trying to walk through quicksand.

Finally, late in the evening, the three of them were seated at the dining room table. The phone had stopped ringing, and they started talking. Partially by nature and partially by the awareness they had gained from The Grief Recovery Institute, they started talking about their relationships with NaNa. Of course, for Rhonda, NaNa was Mom. Jack, too, had started to call her Mom shortly after he and Rhonda married.

As they talked about the woman they had all adored, they laughed, they cried, and they remembered. There were several stories that they had in common. And they each had their own special memories of events that had happened over the years.

At one point, Jack said, "Even though I always told her I loved her, as we've talked, I have been realizing that there are several things I never told her."

Prompted by Jack's comment, Amanda and Rhonda each began to think about some things that they'd also never said to NaNa.

As the natural relationship review continued, Rhonda remembered some of the ideas that John and Russell had reminded her about that day on the phone. She and Jack were able to use the ideas from the Emotional Energy Checklist to help themselves and Amanda. Amanda had a lot of energy in many of the different categories. That night and over the next few days, Amanda was able to discover many things she wanted and needed to communicate.

Here is the Emotional Energy Checklist about Amanda's relationship with NaNa.

EMOTIONAL ENERGY CHECKLIST

GRANDPARENT, RELATIVE, OR CLOSE ACQUAINTANCE

Again we remind you that this checklist is only a guide. It exists to help you to be able to remind a child of the kinds of events that might have produced emotional energy for the child. Since this list is a composite, it is unlikely that children will have stored energy and undelivered communications in each and every category. In fact, for some children, there may only be a few of these categories that contain anything they might want and need to talk about.

It is also important to use the list as a way of helping children review *their* relationship with the person who died. Again, be careful to help your children discover only what is true for them. You may find it helpful to have this list handy when you talk with your children. There is a section at the end of the list where you can make some notes. For older children, it is acceptable to let them have the list and encourage them to make their own notes.

✓ Meeting or first awareness
✓ Mom's parent or dad's parent (or other relationship)

✓ Special names—Nanny, Poppy, etc.

✓ They took care of child [baby-sitting] or child stayed at
    their house

✓ Punishing or easygoing

✓ Gifts, or lack of gifts, and/or better gifts to siblings or others

✓ Trips to their house

✓ Visits to child's house (kids sometimes lose room to
    grandparents on visits)

__ Smells—alcohol, perfume, medication, tobacco

__ Grandparent fights with Mom or Dad

✓ Very safe and easy to be with or talk to

__ Scary

__ Pinches cheek too hard, teases, embarrasses

✓ Personal idiosyncrasies—positive or negative

__ Lives near

__ See a lot

__ Love the visits—hate the visits

__ Don't see much—happy or sad about that

✓ Lives far away

__ Don't see much

✓ Frequent visits

✓ Lots of phone calls—good or bad

__ Few calls—happy or sad about that

✓ Child observes how grandparents interact with each other

__ Child observes how parent interacts with grandparent

✓ Stay with grandparents when parents on vacation—
    like/don't like

__ Wants to live with grandparent when mad at parents

___ If local—shows up at school and other events (birthdays)
     like/don't like

✓   If distant—calls re: school events/birthdays/ like/don't like

LONG-TERM ILLNESS
        **(This section is not relevant here since Amanda' s
        NaNa died suddenly)**
___ First awareness of illness—reaction

___ Child's observation of parents' reactions

___ Diagnosis, treatment, medications, especially relevant if
     local

___ How does parent talk about their feelings regarding the
     illness?

___ Is child allowed or encouraged to talk to grandparent?

___ Is child willing to talk?

___ Is there anyone else child might talk to?

___ Visits and what happens at visits

___ When it appears that illness is terminal and how is that
     communicated?

___ Is there anyone for the child to talk to about the potential
___   death?

___ Do parents pressure the child to visit or call, even if child is
     unwilling?

___ Near the end

___ Circumstances and events that child remembers from the
     last days

___ Emotional response (or lack of) to those events

___ Was child included in the hospital/hospice care? Did
     child have a choice?

___ Anyone safe for child to talk with about what was happening?

___ Does child try to take care of parents' emotions?

THE LAST DAY—OR SUDDEN DEATH
  (Now we return to Amanda's story.)
___Phone call if far away

✓Who told the child and how?

✓Emotional impact on child, if any

✓Did parents show or express emotion in front of child?

___Child at bedside—home or hospital

✓Last conscious interaction—phone or in person

___If in coma—did child speak anyway?

✓Anyone safe to talk to about the event?

FUNERALS, BURIALS, AND OTHER MEMORIAL SERVICES

Chapter thirty-three discusses the question of young children's attendance at funerals.

Please read it before making a decision on behalf of your child.

CHRONICLING EVENTS THAT OCCUR AFTER A DEATH

We are showing you this category to help you remain aware that your child will continue to have memories about the person who has died. Special events tend to act as a trigger or reminder about people who are no longer here. This is a big category for Amanda. We are not going to go into the details here, but based on what you have read about Amanda, you can imagine that there have been many events at which she was aware that her NaNa was not there.

In the next chapter a section called "New Discoveries" will show you how to help your child deal with some of the emotions generated by reminders of someone who has died.

Holidays, birthdays, any other special days

Recitals, sporting events, art exhibits

Graduations, confirmations, bar mitzvah

Parents fight and divorce (important, if grandparent who was safe to talk to has died)

Future: career, marriage, and children

NOTES:

_____

_____

_____

_____

_____

_____

_____

_____

_____

_____

_____

We have tried to make this Emotional Energy Checklist inclusive enough to open up all of the potential areas where your child might have stored emotional energy about the relationship. Again, you will probably discover that you have emotional energy in many of the categories, as well. Please let that be okay with you.

In the earlier letter that Jessica wrote to Thumper, we showed the recovery categories (A = Apologies, F = Forgiveness, etc.) following each statement in that letter. As you read Amanda's letter you will see how each comment fits in one or more of those recovery categories.

AMANDA'S COMPLETION LETTER TO NANA

Dear NaNa,

I have been remembering our time together and I have found some things I want to say.

These last few weeks since you died have been so difficult for me. Each time I remember that you are no longer here, my heart aches. I have been thinking about all the times I called you or talked to you when I was unhappy about something, and you always listened to me. Now I hurt more than ever, and I really want to talk to you. But you are not here. I miss you.

NaNa, most of the times we had together were wonderful. But, there are a few things that I want to apologize for.

I told my girlfriends that you were "weird" about keeping everything so neat. I'm sorry that I said bad things about you.

I'm sorry for the times when I wasn't nice to you.

I'm sorry for the times I didn't listen to you.

There were a couple of times when I didn't tell you the truth. I'm sorry for that.

NaNa, I forgive you for always being on my case to clean up my room.

I forgive you for the few times when I thought you didn't approve of some of things I said or did.

I forgive you for not liking some of my girlfriends.

I never understood why you were so crazy about keeping

things so neat. At your funeral, I talked with your sister, my great aunt Sylvia, and found out that when you were a little girl, you had been punished for being "messy." I understand a little better now.

You were the best teacher I ever had. Thank you for being so patient with me.

I know that you hated to travel, but you came to visit me so many times that I can't count them all. When I think about that I realize that you loved me enough to do something you hated. Some of the times that I have doubted myself, I have remembered how much you loved me. Thank you for always loving me.

I loved how sweet and considerate you and PaPa were to each other.

NaNa, one of my sweetest memories happened on the last day of your life, when I gave you the sketch I had made of you. I will never forget the look on your face, and the tears on your cheeks as you looked at the drawing.

I think that I thanked you for all the gifts you gave me, but I want to say it again, thank you. Every thing you gave me is precious to me.

NaNa, I thank you for being you and loving me. I am so sad that you are not going to be there for all the things that are going to happen in my life. I will always remember to tell you everything, even though you are not here.

I love you, I miss you,
Good-bye, NaNa

(Again, please notice that Amanda sometimes starts her comments with "NaNa." As we mentioned this can help stimulate the normal emotions attached to loss.)

CHAPTER 25

One More Letter

As you recall, this book began with one mother's phone call: "My son's father died, and I want to know how to help him." We promised you that we would tell you how that situation concluded. The boy's name is Jeffrey. With a little help from The Grief Recovery Institute, Jeffrey's mom was able to help him review his relationship with his dad and write a letter that reflected what he needed to say. What follows is a simple version of that review and the letter Jeffrey wrote to his dad.

Jeffrey and his dad were exceptionally close. Their big bond was hiking. Jeffrey's dad had grown up in the country, and he loved the out-of-doors. Ever since Jeffrey could walk, his dad had taken him hiking and taught him about animals and trees and the great outdoors.

Jeffrey's prize possessions were his hiking boots and his Swiss Army knife that his dad had gotten him for his ninth birthday. Some days Dad would leave work early and surprise Jeffrey by showing up at his school at three o'clock, and off they would go for an unplanned hike. Jeffrey lived for those events. They would find a quiet spot and just sit and talk.

Like most parents and children, they had their occasional rough spots. Dad was very levelheaded and approached life in an orderly fashion. Jeffrey tended to be a little more impulsive. When they clashed, it was usually because of these differences in style. Jeffrey didn't always like to admit to being wrong, even if he was, and he wasn't the best in the world at apologizing.

Some of Jeffrey's schoolmates had some real problems at home, especially with their dads. Jeffrey considered himself a very lucky boy to have such a wonderful dad.

JEFFREY'S LETTER

Dear Daddy,
I have been remembering our time together and I have found some things I want to say.
Daddy, I miss you.
I'm sorry for all the times I was stubborn.
I'm sorry for arguing with you.
Sometimes I thought you were too strict with me, I forgive you.
You never let me win any arguments, I forgive you.
Daddy, sometimes I just stare off into the hills, and think about the times we went hiking.
I want to keep my Swiss Army knife forever, because it reminds me of you.
I think you were the smartest person in the world—even though it made you win all the arguments.
I was really proud of you. I don't think I ever told you that. I'm sorry. I hope you knew how proud I was.
Daddy, I'm so sad that you had to die, it's so unfair.
I love you, I miss you,
Good-bye, Daddy

You might notice a substantial difference between Jeffrey's letter and the letters of Jessica and Amanda. Jeffrey was nine years old when he wrote the letter. It reflects his individual communication level and the fact that he was nine. Jessica and Amanda were thirteen and fourteen. Their letters represent their relationships, their

ages, and their communication skills. Remember, we never compare any aspects of grief, recovery, or completion.

Each of the letters, as well as the verbal communication of the
four-year-old to Mr. Hamster, is accurate for the individual who
wrote it. As children mature, they may have new thoughts and
feelings about relationships that ended or changed sometime in the
past.

NEW DISCOVERIES

The relationship reviews and letters help children become emotionally complete with all of the feelings, thoughts, and ideas they
remember about the person, the animal, or the event that has
affected their lives. Children will be able to discover and complete
those things they become consciously aware of as they review the
relationship. The actions we have suggested will assist children in
uncovering a tremendous amount of emotional connections. Having done so will make it possible for them to discover other things
in response to future reminders about those relationships.

Each new discovery can be completed in the same way we have
already shown. That is, with apologies, forgiveness, significant
emotional statements, and fond memories. However, there is no
need to start all over each time a new discovery is made. A helpful
device is simply to write a P.S. letter. For example, Jeffrey might be
at a ball game and remember that his dad had told him about the
heroes he had looked up to when he was a kid. Names like Stan
Musial, Mickey Mantle, and Willie Mays might come into Jeffrey's
mind. Jeffrey might have a sweet and sad moment thinking about
the times that his dad had told him how great those players were.
Later, Jeffrey might sit down at his desk and write the following:

Dear Daddy,
I was at a baseball game today, and I remembered all the times
you would tell me about all the great players you watched when

you were a kid. I don't know if I ever told you how much I loved listening to you talk about those times. I really did enjoy that. Being at the game without you was sad. I miss you.

  I love you,

  Good-bye, Daddy.

It would be a very good idea for Jeffrey to read that P.S. letter to his mom or to someone else whom he trusts. Writing the letter and saying it out loud helps complete that undelivered communication, and the process allows Jeffrey to stay current with the ongoing emotions that relate to the absence of his dad. By taking this action, Jeffrey retains the freedom to remember his dad with any or all emotions attached. He does not have to be afraid to think about or talk about his dad. As Jeffrey grows and matures, he will always be able to communicate his thoughts and feelings.

WHAT ABOUT JEFFREY'S SISTERS?

Jeffrey's fourteen-year-old sister was the one who had decided to save the family and was busy being strong for everyone. Our first task was to help her mom understand, for herself, that being strong was not a good idea. Then we guided her in helping her daughter make a review of her relationship with her dad. She then wrote her own unique letter to her dad.

  The little sister, who was five, was the one who had copied Mom's busyness and had become a whirling dervish. For her, we helped Mom to just have some little chats where she told the truth about what she was feeling. Simple comments like, "I miss your daddy so much." Pretty soon the little girl began to say the things that she was feeling and the things she missed about her daddy. Instead of writing a letter, her mom helped the five-year-old say some of those things to a picture of her daddy. Now she knows how to do that, and how to finish every time with "I love you. Good-bye, Daddy."

CONCLUDING PART FOUR

The action of reviewing a relationship is natural, but it is also help-ful to become aware of the areas of relationships that are most likely to produce emotions. The essence of a relationship is con-tained in a child's emotional attachments, whether to people, to animals, or even to places and things. Completion is the result of the actions which discover and communicate everything that had emotional value for the child.

> The key to completion is that the thoughts, feelings, and
> ideas must be verbalized and be heard by another living
> human being to be a "completed" communication.

To illustrate the importance of this fact, we will tell you a story about a woman who came to one of our personal workshops years ago. Her husband of forty-two years had died. Theirs had been a wonderful marriage, which included all the normal ups and downs that are part of a glorious relationship. Before attending the work-shop, she had done almost all of things that we ask people to do, including writing a letter that contained apologies and forgiveness. She then went and read the letter at her husband's grave. And she cried.

For a few days she felt better. Then she started feeling very low. She couldn't understand why the letter had not released her and allowed her to move on. Some time later, at the end of her emo-tional rope, she showed up at a Grief Recovery Personal Work-shop. At each juncture, when we asked her to take one of the actions that lead to recovery, she said she'd already done that. We asked her to please do it again. We did not know what she had done or not done; we knew only that she had shown up at our workshop, so she must have missed something. We just didn't know what. On the last day, in the last few hours of the workshop, we made the statement we always make at that point:

> Undelivered communications of an emotional nature must
> almost always be verbalized and must almost always be heard
> by another living human being to be a "completed"
> communication.

There was a gasp in the room. The woman realized what had happened, or, in her case, what hadn't happened. She had read her letter at the grave, which sounds so wonderful, so right. But it had not been heard by another living human being.

She then read her letter to the partners in her group, and she cried. And she felt better. She still feels better today, some eight years later. She is sad from time to time and misses him, but the pain is gone.

Children need to be allowed, encouraged, and assisted in discovering and communicating what is emotionally incomplete for them. The actions of doing the relationship review, making the conversion into recovery components, and writing the completion letter can be tremendously helpful. But the most important action is that they have the opportunity to say those emotionally important things out loud, and to be heard by safe and accepting hearts and ears.

The Grief Recovery Completion Letter is an ideal vehicle for delivering undelivered communications, and it works especially well with children who can communicate their thoughts and feelings in writing. But, as you remember the four-year-old, who would not have been able to write a letter, what was most essential was that he said his emotional comments out loud, and that someone else heard him.

When you listen to your children say or read their good-bye thoughts and feelings, you must adopt the image of being a heart with ears.

Listen with your heart.

FOCUSING ON MOVING AND DIVORCE

Very early in this book we mentioned that there are more than forty life experiences that can produce the conflicting feelings called grief. These experiences, and how children accommodate them, can be the source of tremendous emotional and physical damage. We have prepared an edited version of this list to include only those losses that are directly applicable to children. As you will see, it is a formidable collection. You will also notice how many of the loss events use the word *change* to define them. Remember our definition: Grief is the conflicting feelings caused by a change or an end in a familiar pattern of behavior.

Death of close family member

Divorce of parents

Moving

Death of a close friend

Death of a pet

End of romantic relationship (teens)

Personal injury or illness

Change in health of family member

Gain of new family member

Recognition of outstanding personal achievements

Beginning or end of school

Change in schools

Change in social activities

Change in eating habits

Vacations

Christmas, birthdays, other holidays (particularly the first
    following a death or divorce)

In part four we focused on the death of a pet and the death of a
relative, and on the actions that lead to completion of those kinds of
painful events. Now we are going to look at non-death losses. One
of the most overlooked of all major losses is moving. The first section of part five shows the right way to guide a child through this
change; the second section shows the wrong way. We believe you
will find this to be very enlightening.

The balance of this part is devoted to divorce. We hope that you
never need this information, but, if you do, it will be very helpful.
If you are not involved in a divorce, read it anyway. It may come in
handy when and if one of your friends and his or her children are
caught in the middle of a divorce.

CHAPTER 26

The First Big Move

In 1987, John, his wife, Jess, and their six-year-old son, Cole, were preparing to move from an apartment in one area of Los Angeles to a house in a new neighborhood. By this time, John had been helping grieving people for many years, and he knew that grief can be defined as the conflicting feelings caused by a change or an end in a familiar pattern of behavior.

John had long since recognized that the first move from one home to another is one of the most powerful loss experiences affecting children. He knew that it did not matter if the new house or apartment was bigger and nicer than the old one. He also knew that it did not matter if the move was from one city or state to another, or simply from one part of town to another.

Children often struggle with change, for change can be scary. Moving automatically represents changes in everything that is familiar to a child. Guess who is also affected? If you said the parents, you would be right.

Often when there is a move, it can mean that there is more money and the move is to a larger house. Those are positive things. However, regardless of its size or condition, children are accustomed to the old place. They know the home, and it seems to know them. They know every nook and cranny, for it is their home. The exciting feelings about the new home are mixed together with sadness about leaving the old one. Even if the children did not love the old home, they were very familiar with it. That mixture of positive and negative feelings illustrates what we mean by the phrase "conflicting feelings."

Sometimes fortunes are reversed, and the move is from a larger house to a smaller one. This move not only represents a change in the familiar but also adds the negative feelings associated with financial difficulties. Although young children may not recognize or relate to the money problems, they will be affected by their parents' attitudes. Children often hear parents' late-night arguments about money. Or they are aware of their parents' nonverbal communications, which indicate that something is not right.

> It is wise to remember that all major changes create
> emotional energy in children and adults.

Cole was excited about having a house with a yard and his own swimming pool. He was also excited about having a larger room. At the same time, Cole was sad to be moving away from the friends he had made at his old school and in his neighborhood. John knew that the move they were about to make was a golden opportunity to teach Cole how to deal with the confusing feelings he was experiencing.

John took his family on an emotional tour of their apartment, and Cole very quickly got into the spirit of the occasion. They talked about the happy and sad experiences they had shared in each room. They thanked each room for keeping them safe and protecting them from the hot or cold weather. They remembered important things like when Cole had lost his first tooth, and when he had first learned to write his own name. And as they left each room, they said "thank you" and "good-bye" to it.

This exercise wasn't just for Cole's benefit. John and Jess were able to remember and talk about many of their memories, both good and bad. The process was very helpful to all three of them. On moving day, with tears in his eyes, Cole waved good-bye to the only home he had ever known. Cole adapted very well to the new home. Having completed his relationship with the old apartment, he was able to develop a new relationship with his new house. John and Jess still have fond memories about the old apartment in which Cole spent his early years and they have built many more wonderful memories in their new home.

Soon Cole will go away to college. While his room at home will still be his on holidays and vacations, he will take the same actions he did thirteen years ago, recognizing that many of his day-to-day actions will change as the result of his move. John and Jess will be with Cole, and together they will remember the events of the past thirteen years. Most of his life will now be lived at his college dorm, where he will develop new familiar patterns of behavior. And you have probably already figured out what Cole will need to do four years from now.

TRANSITIONAL EVENTS

We have seen countless examples of the negative effect on children (and adults) who do not properly complete their relationships to people and things from the past. This simple exercise can help ensure smooth transitions. Moving is just one of those transitions.

There are many other chronicling events where this same kind of completion can be life-affirming. Do you remember your first child's first day of school? We'd bet you do. You are probably beginning to think of things like awards, recitals, and graduations, each of which can represent a change to a new level of familiarity. These events represent golden opportunities for you to help your child review the events and emotions that have led to that point and to move on to the next events in his or her life.

Even if this exercise seems silly to you, we cannot overemphasize its importance.

CHAPTER 27

What Not to Do

The previous story about Cole's big move illustrates the most effective method of dealing with moving. However, it may help you to read a story in which everything was done wrong, and to discover some of the consequences of this mishandling.

A young man named Tommy was twenty-eight years old when he participated in one of our seminars. He related a story that we have heard many times. His tale is so descriptive of the potential problems that can result when moving is not handled correctly that we include it here as a negative example.

Until he was eight years old, Tommy lived happily in a nice neighborhood in a city in the Midwest. He attended a school just a few blocks from his house. Several of his friends lived on his block, and he and his pals did nearly everything together.

Tommy's dad worked for a large company that decided to transfer him to an office in the Pacific Northwest. The new job included a substantial raise and a higher job title.

Tommy remembers his dad and mom telling him that they were going to move. He remembers being very upset with the idea of moving away from his friends, his school, his house, and all of the familiar places in his community. Through his tears, he told his parents that he did not want to leave his friends. His dad said, "Don't feel sad, you'll make new friends." We're sure the phrase "don't feel sad" looks familiar to you. Remember, earlier in the book we mentioned that almost every incorrect communication about emotions begins with "Don't Feel Bad" or "Don't Feel Sad." Keep in mind that dads and moms are important sources of information and guidance for children.

While it is a reasonable idea that we can make new friends once we move, it is unreasonable to say that we shouldn't feel bad about leaving the friends we do have. Even worse, children who are not helped to acknowledge and complete the pain caused by the separation from their friends often will not make new friendships. Children who move several times during their childhood tend to become observers rather than participants in their own lives. Why make new friends? becomes a very logical internal question. While the moves themselves are important, it is the emotions that are caused by the moves that can create long-term problems.

Tommy was confused. He thought that perhaps his dad had not heard or understood him. He did not understand why he should not feel sad about leaving his friends. He tried again. This time he told his parents that he didn't want to leave his school. Once again he was told, "Don't feel sad, you're going to go to a better school." Tommy had a very hard time with this idea. He loved his school, and he loved his teachers. He couldn't imagine leaving all of that behind. And besides, he couldn't understand why his parents were telling him to not feel sad. His whole life they had encouraged him to tell the truth. Now when he told the truth about how he felt, they told him he shouldn't feel that way.

Tommy tried one more time. He told his parents that he didn't want to move from their house, and particularly his room, which he had spent so much time fixing up with baseball posters. Once again he was told, "Don't feel sad, we're going to have a bigger house, and you'll have a bigger room."

Using baseball language, Tommy was now 0 for 3. He tried three times to get his parents to hear him, and each time he was told "Don't feel sad," followed by "new friends," "better school," and "bigger house." Finally, Tommy quit trying. Each time he had voiced his emotional concern about his friends, his school, and his room, his parents told him not to feel what he was feeling and then gave him an intellectual reason why he shouldn't feel sad.

After the family moved, Tommy, who had been a very outgoing boy, began to turn inward. He did not develop new friends after the move. He did not do well in school. And he never put any time or energy into fixing up his new room. His story is typical of the kind of painful stories that we hear all too often at The Grief Recovery Institute. Fortunately, Tommy came to one of our seminars in his twenties and was able to turn his life around.

The tragedy in this story is that Tommy's dad missed out on a golden opportunity to communicate something very helpful to his son. Tommy's dad had been born and raised in that same Midwestern town from which they were going to move. Tommy's dad had spent his whole life there. All of *his* friends were there. Tommy's dad also did not want to leave everyone and everything he knew.

But, he did not tell the emotional truth to his son.

It would be unfair to say that he lied to his son, but it is accurate to say that he did not tell the truth. Each time Tommy's dad told Tommy that he shouldn't feel sad or bad, he moved Tommy just a little bit further from the truth, and a little bit further from being able to trust his own father's advice.

Tommy's father was not a bad dad; he was simply passing on to his son the same misinformation that had been passed on to him when he was a child.

By now, you probably have figured out a better way for Tommy's dad to have helped his son and himself. In response to Tommy's very first comment about not wanting to leave his friends, his dad might have said, "Me, too! I'm afraid to leave all of the friends and family I have known my whole life. Even though I'm excited about my new job, and more money, and the great adventure that lies ahead, I am sad and a little afraid of leaving everything I know."

Earlier we defined grief as the conflicting feelings caused by a change or an end in a familiar pattern of behavior. Dad's comments above clearly illustrate conflicting feelings. Sad and scary feelings about leaving an entire lifetime behind are entangled with exciting and happy ones about a better job and great new adventures.

And Dad might have taken Tommy to visit his friends to tell them that he was moving and that he was sad about leaving them. They would have made sure that they got everybody's phone number and address. And after arriving in their new home, Dad would have made sure to help Tommy stay in touch with his old friends. And that would have reminded Dad to stay in touch with his friends, too. And so Tommy's life might have worked out differently.

We hope you can see the difference between the two scenarios. We have seen lifelong negative consequences when the process of moving is handled poorly. And it's so easy to do it well. Here are a few suggestions to guide you when talking about a move:

MOVING

HAVE A PLAN: It is not a good idea to barge into the circumstances of a move unprepared. It is much better to develop a plan for dealing with the feelings that more than likely will come up for your child.

TELL THE EMOTIONAL TRUTH ABOUT YOURSELF: Since the parents are the emotional leaders of the family, it is always a good idea for the parents to lead the way. A simple emotional statement will work, like, "I'm really excited about the new house, but I am a little sad about leaving the place we have lived in." If that isn't true, tell the truth. "I'm excited about the new house, and I never liked the old one. I did get used to it, but I never liked it." As you can see, that is a truthful statement, but is also an emotional one. What we want you to do is make it safe for children to tell the emotional truth.

WHEN POSSIBLE ADD A FIRST-PERSON STORY: If you have a recollection of a story from your childhood, tell it. Again, use the story to create safety for the naturally occurring emotions that attach to all major changes.

ALL FEELINGS ARE OKAY AND NORMAL: Remember that love and hate, and happy and sad are all normal feelings. Don't create a hierarchy of feelings. And whatever you do, do not indicate that sad, painful, or negative feelings are bad.

GETTING PHONE NUMBERS AND ADDRESSES OF FRIENDS: Be real careful not to make any empty promises. If you tell your children that you are going to help them gather names and addresses of friends, do it. It is also important that you help your children remember to write the letters or make the calls, very soon after the move.

REALISTIC COMMENTS ABOUT FUTURE VISITS: This can be a very troublesome area. We can't tell you how many people who have come to our workshops were still affected, often decades later, by their parents' not following through on promises to visit the old hometown. Often in an attempt to deal with children who are very emotional about moving, parents will promise visits. Children will take those words literally.

CHAPTER 28

On Divorce

LESLIE GETS THE FIRST WORD—
THE DIVORCE OF MY PARENTS

One of the most painful and significant events in my life was my parents' divorce. It felt as if my family had died and my whole life as I knew it would change forever. For me, the divorce was **sudden and unexpected.** My father left the house one night and would never return as the Dad I grew up with. He had fallen in love with someone else, which shattered all my thoughts and feelings about what a happy marriage my parents had. **They were supposed to be together forever** . . . "until death do us part." My body literally went into shock, and I cannot recollect what I did for days or weeks after hearing the news about my parent's separation. I was about to graduate from high school, and the last few weeks of school were a daze.

All I could think about was when would my dad return to my mother and this whole nightmare be over. I had my high school prom to go to, which I hated and did not want to attend. I don't even remember if any family member said good-bye to me as I embarked on what was supposed to be a night to remember.

I remember my father driving to the house for visits with my mother. My siblings and I would wait to see him, and sometimes he would leave the house without ever saying hello or good-bye. None of us could really be there for each other because we were all hurting over a relationship that was unique and special to each one.

I immediately stepped into the role of caretaker, especially with my younger siblings. My mother could not function as a mom at that time; her pain was so tremendous that I think for her to get up and even join us for dinner was amazing to us. **Everything that was familiar was gone.** We had our family dinners every night, and I waited for my father to walk through our door every night. Even our dog, Poof, wouldn't eat, and the vet told my mother that our dog was depressed and was having a breakdown.

I wanted to fix the family; I wanted to make the pain go away. I was scared to tell my father how I really felt or how angry I was about the divorce, because deep down **I was scared I would lose him even more than I already had.** I wanted to please him and accept the situation no matter how painful it was to my family or me.

I turned to food as a comfort. I had never before had an issue with food. I was healthy; I loved sports; and my body was in good shape. My first year away at college, I developed the eating disorder bulimia and used food both as a comfort and as a punishment. I gained a lot of weight the first year after my parents' divorce, and my eating disorder consumed all my free time. I really had no time for dating or being with my friends. On weekends I would come home and have my worst binge and purge sessions with food, sometimes vomiting twelve times a day. For approximately two years I struggled with this disorder, which did nothing to help with the pain I was feeling from my life and from everything I felt I had lost. I then went in the opposite direction and began to get obsessed with losing the weight and exercising like crazy. I would drive forty-five minutes to get to an exercise class that started at six in the morning. **I went into "busy" mode** so I would not have time to think about what was painful in my life. My relationship with each of my parents was changed forever, and my heart remained broken.

Years later with the help of therapy, I really wanted to have a life that I could live to the fullest. I understood there were things I could not change, and I had given up on the hopes and dreams of having my family back the way I remembered. My relationship with my future husband was going wonderfully while we were dating—until he proposed to me. I was very much in love, and yet when Brian proposed, I realized that I had everything to lose. **I tried to sabotage the relationship.** It wasn't a well-thought-out plan, just a plan that occurred out of fear. I did not want to be a statistic of divorce. Luckily, Brian is a wonderful man who stood by me, and we worked together on the issues of marriage. We have been married for ten years, and we've just had our third child.

It was shortly after my dad's death that I showed up at the Grief Recovery Personal Workshop. Although it appeared that my reason for being there was the death of my father, Russell and John helped me become aware that I was still incomplete with my parents' divorce all those years ago.

We have highlighted several of the comments in Leslie's story. As you think back on the beginning of this book, and as this section on divorce continues, you will be able to see that many of the elements of Leslie's story match up to the concepts we are discussing. Again, we want you to realize that we are talking about reality here, not theory.

CHAPTER 29

Bad New—Bad News

We have bad news, and we have bad news. That is not a misprint. Throughout this book we have encouraged you to help your children discover what is emotionally unique and specific in their relationships. We have been cautioning you to be careful to avoid confusing your relationships with your children's relationships. Unfortunately, much of it will not be applicable if you and your spouse are divorcing or have divorced.

It would be naive of you or us to believe that you could maintain any semblance of detachment from your own thoughts and feelings about your own divorce.

We ourselves are very good at dealing with our own emotions, and, frankly, we wouldn't try it. In fact, when any of our family members participate in Grief Recovery Personal Workshops, they do not attend those that are led by their parent or spouse.

We agree with the idea that a lawyer who represents himself has a fool for a client, or that doctors shouldn't doctor their own families.

LONG TERM OR SUDDEN IMPACT

In our certification training seminars we teach that there are essentially two different categories that relate to death. One is long-term illness, and the other is sudden death. In this book, we have high-lighted that distinction by separating these categories in the Emotional Energy Checklists. Long-term illness and sudden death produce different areas of emotional energy in those who are struggling after the death has occurred.

It might surprise you to note that we divide divorce in very much the same way, that is, long-term or sudden. The difference with divorce is that there is often one partner who has been struggling for a long time, while the other partner has been unaware that things are not right. When the latter gets served with divorce papers, it can have the impact of a sudden death.

Some children are very aware of a problem in their household. They have often seen and been subjected to arguments between their parents over an extended period of time. For those children, the announcement of a divorce will fall under the heading of a long-term condition. On the other hand, some parents manage to conceal from their children their personal difficulties with each other. When children who were not aware of any major problems are informed of an impending divorce, their reaction is also as if a sudden death has happened. The impact can be overwhelming to a child. There is a high probability that a child may begin to participate in a variety of short-term energy relieving behaviors, in response to the sudden news of their parents' divorce.

Before we make a few suggestions as to how your children can benefit from what you are learning in this book, we want to discuss a few important ideas.

WHOSE DIVORCE IS IT?

You might recall our questions at the beginning of the book: What's the problem, and Whose problem is it? Here we go again. The question now becomes, Who is getting the divorce? It could be said that divorce is a family matter. And even though there is a truth in that comment, the bottom line is that the couple is getting the divorce, and the children are in the line of fire. The collateral damage to the children can be monumental.

MULTIPLE LOSSES

The children caught in a divorce are experiencing multiple losses. What loss or losses are they experiencing? Let's define loss again as the conflicting feelings caused by a change or an end in a familiar pattern of behavior. We know we are being repetitive with that definition, but we really want you to understand it.

Among the losses for the child are:

> Loss of expectation that this family would be together
>
> Loss of trust
>
> Loss of familiarity and routines
>
> Loss of safety
>
> Loss of childhood
>
> Loss of residence, and/or the change to dual residences

Any one of those losses is enough to break a child's heart. Taken in concert, they are overwhelming. Let's look at each in a little more detail.

Loss of expectation that this family would be together. Children are taught about love and honor and trust and loyalty by their parents. They learn how to be loving and considerate, how to resolve conflicts, and how to get along with others. From literature and films and religious institutions, children learn that the vows exchanged in the marriage ceremony pledge a commitment to those virtues. Whether or not you've experienced this, think about how confusing and disturbing it must be to children when their parents cannot maintain that pledge to each other.

Loss of trust. Imagine the conflicting feelings children must experience as a divorce scenario unfolds or explodes before their eyes. What reference point do they have to deal with those feelings? It is very difficult to teach your children about love and simultaneously teach them about divorce. Given the implicit promise that the family will always be together, the divorce itself represents a major breach of trust.

Loss of familiarity and routines. This is difficult all by itself, and it's often greatly intensified by the fact that children may be undergoing other major transitions as they move from childhood to adulthood. We know all too well that the stresses and strains of those transitions can have powerful consequences. Those transitions can be happening in every age bracket.

Loss of safety. Familiarity and routines build safety and a sense of well-being. The patterns established within a family are usually dismantled by divorce. Children, flailing around in the emotional aftermath of a divorce, often do not feel very safe. Safety and familiarity go hand in hand, so it is a good idea to limit the amount of additional changes.

Loss of childhood. The instinct for survival can take many forms. For the most part, survival actions are beneficial. Sometimes, though, they backfire. The scenario in which children take care of a parent is one example of such a backfire. It is understandable that children would instinctively try to protect the very person or people who are supposed to protect them. It's the children's way of trying to guarantee their own survival. But this impulse to "caretake" puts them in conflict with their own nature.

Divorce tends to turn children into amateur psychologists. It spurs them to analyze and figure things out. It forces them to grow up before their time and to take on attitudes and actions that are not appropriate to their time of life.

Loss of residence, and/or the change to dual residences. Part five opened with a discussion about moving. Everything we talked about in that section is magnified when the move is the result of divorce. The moves or changes caused by the divorce carry an almost incalculable emotional weight, which is added to the fact that moving, in and of itself, changes everything that is familiar and routine.

SOMETIMES WE GET LUCKY

As we approached the writing of the final chapters of this book, we realized that it would be a good idea to have some first-person stories to illustrate the benefits of the actions of grief recovery on the lives of children. We called a few friends and asked them to write up some of the life experiences that demonstrate how their own participation in grief recovery has helped their children. The response was overwhelming. You will see the results in part six, in the Win-Win section.

One of the stories we received is so relevant to the issues about dealing with the day-to-day issues of divorce that we have included it in this section. You will notice some language in this story that is identical or parallel to ours, which is why we say we got lucky. This story reflects in perfect, real-life detail our point in this chapter. You will notice that the story names five of the losses we have just reviewed.

From our friend, Jeff Zhorne:

The benefits of grief recovery to my three children are enormous. Perhaps my oldest natural child, Sophia, seven, gains the most from the help and healing of grief recovery because she has to cope with split households. My former spouse and I have a suitable custody arrangement, but it still comes down to packing for Dad's every weekend, and then school evenings with Mom.

In a split residential situation, children experience a number of losses, including loss of trust, loss of familiarity, loss of residence, perhaps even loss of safety, among others. When she is away from one parent or ending an extended stay, Sophia often expresses feelings of sorrow about missing her mom or me. During those times I feel I can be emotionally present with her as she expresses those feelings of being torn or wishing we were both still together.

As six o'clock approaches every Sunday night, as it always does, signaling the end of another restful yet playful weekend at Dad's, Sophia often sits dejectedly and quietly in the backseat as we make our way to Mom's. During those times I recall that, thanks to grief recovery, I can try to provide a safe environment for her to express any sad, painful, or negative feelings. Most of the time she does give voice to those emotions.

And that is when my years of effort to be emotionally complete with the losses in my own life pay off: I can simply be present with Sophia as she bemoans her state. I can simply *be*, not *do* anything. I don't have to fix her, give advice, or jump in and change the subject because she may have triggered some latent unresolved issue of my own. I can listen, hear, and acknowledge Sophia as she trusts me with the very specific feelings she is experiencing about starting to miss the parent she's about to be away from for a week. Grief recovery is most helpful because I can simply connect with Sophia at an emotional level, whatever that happens to be, and not take away her feelings.

Frankly, we are overwhelmed with gratitude and pride, knowing that what we teach actually works. If you read Jeff and Sophia's story carefully, you will realize that the key to the success is the work that Jeff has done on his own loss issues. We cannot emphasize enough how important it is for you to do the same. Get a copy of *The Grief Recovery Handbook* and get to work. Your children will benefit exponentially.

DON'T FIX FEELINGS

You might remember that, after John's son died, John kept scouring the bookstores, looking for a book that would help him deal with the emotional pain he was experiencing. You may also recall that John kept finding books that described how he felt. But John already knew how he felt. What he needed to know was what to do about it.

With this in mind, we are not going to try to tell you how your children might be feeling in response to your divorce. If there are twenty million children currently living in and around the circumstances of their parents' divorces, then there are twenty million broken hearts and twenty million different sets of emotions, each and every one of which is accurate.

The biggest problem with sad, painful, or negative emotions is that everybody tries to fix them. Especially parents. Rarely does anyone ever try to fix your happy or positive emotions. You may recall that idea from early in this book when we showed how silly it would sound if someone said "don't feel good" in response to something positive. Feeling good is not a problem.

What we must understand is that feeling bad is also not a problem. Feeling bad just *is*. And feeling good just *is*. Feelings are just feelings. We must stop trying to fix them.

This becomes very relevant when our children are dealing with their reaction to their parents' divorce. This is the time when the parent is most likely to swing into "fix the problem" mode. They will try to fix their children's feelings, which they can't. More important, they will be ignoring their own. It is where the parent reverts to old ideas such as "Be strong for others" or "Be strong for the children."

Remember Jeff's comment in his letter about being able to hear Sophia without having to fix her feelings. The ongoing safety of Jeff not fixing Sophia's feelings allows the door to stay open, and Sophia can continue to trust her dad and trust her own feelings. In the imperfect world where divorce is a reality, you, the parent, need to take the actions that will help you complete your own emotional relationship to your divorce so that you can hear and help your children.

DON'T BE FOOLED—RELIEF IS
ONLY THE LAST FEELING

By the time a marriage has reached the breaking point that results in divorce, many things have usually happened. The lead up to divorce is often years long. Sometimes a great deal of the buildup simmers underground. Sometimes it's all out in the open. Sometimes the children, regardless of their ages, have very little or no idea that there are troubles in paradise. Sometimes they know or have a sense that something is wrong, but they don't know exactly what it is.

For the couple themselves, following the emotional anguish of the long struggle, the actual divorce may come as a relief. A relief from the fighting and bickering. A relief from the often false hope that anything is going to change.

For those children who had a conscious awareness of the problems their parents were having, there may also be some relief. Very likely the parents weren't always easy to deal with after some of their skirmishes.

When children report a sense of relief that their parents have finally divorced, that feeling represents only the end of the hostilities. It does not mean that the child, any more than either parent, is emotionally complete with the hopes and expectations of living in a happy family. It is just the last feeling in a long litany of feelings that have not been comfortable to live with.

NOBLE SENTIMENTS, BUT
HEARTS ARE STILL BROKEN

There was a time when the bonds of marriage were often held together by the concept that it was in the "best interests of the children." As noble as that sentiment may have been, it was probably not the best idea in the world.

For the past thirty years, with a divorce rate that has escalated to near 50 percent, we have witnessed people guided by the idea that it is better to get the divorce than to subject the children to the constant bickering and unhappiness of a bad marriage. Again, a noble sentiment, but probably not the best idea in the world either.

At The Grief Recovery Institute, we have spent the better part of the past twenty-five years helping the children who are the products of both of those philosophies. Neither philosophy precluded the inevitable breaking of trust, safety, and confidence. And neither of them, in any way, limited the emotional pain felt by the children. Everybody loses, especially the kids.

We believe that it is almost impossible to do divorce right, that is, to do it in such a way that it does not affect the children. The parents are the ones who teach the children to love, and then they are unable to sustain the love in their marriage. As an adult, you have to try to imagine the distress and confusion this reality might create in the child's mind.

We are not Pollyannaish. We are not suggesting that there exists an idyllic world in which love is perfect and people are perfect and there is no hate and there is no divorce. We are not sitting in moral or emotional judgment of people who have found themselves in a relationship they cannot sustain.

We are certain, as is everyone else, that hindsight is twenty-twenty. We know that people marry for many reasons, not all of which can stand close scrutiny. We know that people are impulsive. We know that people often bypass their own intuitive senses. We know that there are circumstances that can look okay one day and can seem wrong a short time later.

Knowing all of this does not solve the problem. Children are still affected. No matter what the reason for the divorce, their hearts are still broken.

ONE CENTRAL ISSUE

For us, it all comes back to one central issue. What tools, ideas, concepts, and beliefs do the parents hold about dealing with loss? And what have they, in turn, taught their children about dealing with loss?

In a perfect world there would be no untimely deaths. In a perfect world there would be no divorce. In a perfect world there would be no abuse, no neglect, no abandonment. In a perfect world there would be no disappointments. There would never be rain on a day when someone planned a picnic.

But in this real world, loss is a normal part of day-to-day existence. Obviously, major losses, such as death of a loved one, do not occur every day. But each day may contain some disappointments or disagreements. Avoidance of loss, and the impact it has on our emotions, does not enhance our lives. It is our ability to deal effectively with all manner of loss that allows us to retake a productive and happy place in our own lives.

UNIQUE IS STILL THE BOTTOM LINE

The bottom line is that divorce is going to cause massive disruption in children's lives. Some of the disruption may occur immediately and be very apparent, and some may lie underground, festering, only to rise and affect children later, even after they have become adults. Statistically, there is a high probability that children whose parents have divorced will also experience divorce. We believe that instilling effective ideas about dealing with loss will go a long way toward reversing that trend.

Then again, there are children who are products of divorced households whose lives work out beautifully. Many of them benefit, indirectly, from the divorce, and they use the experience to guide them to making better decisions. We must be careful not to imply that divorce, in and of itself, is bad. It is more accurate to say that each individual child's response to and guidance through the divorce is the key to that child's future.

In everything we do we mention that all relationships are unique, which is why recovery from loss is distinctive for each individual. By the same token, each individual child's reaction to the divorce of his or her parents is specific to that child. We cannot and should not make any absolute statements, lest we make the mistake of painting every child with the same brush. We have seen families where some of the children were devastated by the parent's divorce, with lifelong negative consequences. We have also seen children from the same home who were only minimally affected, both at the time of the divorce and throughout their lives.

In the main, however, it is relatively rare that children are not in some way impacted by divorce. The degree to which they are affected depends on their own sensitivity to emotions of any kind.

As we have discussed, even within a family, children are different from one another. Some are very sensitive and some are thick-skinned. A typical reaction to divorce is often an increase in children's participation in some of the short-term energy-relieving behaviors they may have already developed before the divorce. We know of a family in which one child got heavily involved with drugs and alcohol, one developed an eating disorder, and the two younger children had real "acting out" problems at home and school, all in direct response to their parents' divorce.

Another crucial element is the level of knowledge about dealing with sad, painful, or negative emotions that existed among the family members, adult and child, before the divorce. While knowledge about dealing with grief does not make events less painful, it can lead to actions that can complete what is emotionally unfinished and, in turn, effectively reduce the pain.

WHERE IS THE FOCUS?

While every parent will want to devote attention to their children's well-being, it is very difficult to separate yourself into neat little compartments. It is almost impossible to detach yourself from your own reaction to divorce or pending divorce so that you can focus on your children.

Earlier in this book we gave you guidelines for helping your children deal with a variety of losses: moving, the death of a pet, the death of a relative or friend. We pointed out that you would likely have some emotional energy in some of those relationships. We suggested ways for you to be able to make sure that you did not insert yourself into your children's relationship with the person or animal who died. That was all very important, and we certainly hope that you took it seriously.

When dealing with your children's reaction to your divorce from their other parent, the rules will have to change. It is almost impossible for you or your spouse to maintain anything remotely like objectivity when talking with your children about *your* divorce. Remember, your divorce essentially concerns your relationship with the other parent. Believing that you could have a helpful and meaningful conversation with your children about that relationship is unrealistic. Also, remember that earlier in the book we talked about the danger of turning your children into your confidant or your therapist. We have seen more travesties committed, unwittingly, when parents use their children as a sounding board for their emotions about their former spouse.

It is especially difficult for the custodial parent, who is often overwhelmed with the day-to-day responsibilities of guardianship. It is easy enough to let things slip out, when one is frustrated with all of life's problems. There is a fine line between telling the truth about how you feel and unwittingly encouraging your children to take care of you.

TAKING SIDES

There are several elements that seem almost universal when it comes to children and divorce. First, children somehow wind up thinking, feeling, or believing that they have to take sides. This is simply a matter of human nature. The problem is often enhanced when the custodial parent has the larger responsibility for discipline and behavior, while the noncustodial parent brings gifts and takes the kids on fun outings. Pigeons tend to gather around the person who gives out the bread crumbs.

Children are in their own survival struggle, both within and separate from your divorce. They are trying to learn how to negotiate their own path in life. This is particularly true with teenagers. For those of you who have watched your friends with solid marriages grapple with unruly teenagers, you will know what we mean. The teen years can become infinitely more complicated when a divorce gets added to the struggles of growing up.

CHILDREN SOMETIMES BLAME THEMSELVES

Another difficult element is the fact that, even though parents or helpful relatives and friends assure the children that it's not their fault, children will still tend to blame themselves. Keep in mind that they perceive their very survival to be threatened, and their minds will try to convince them that if only they had done something different or better, their parents would still be together. That children think they caused the divorce might look self-centered to you, but these notions are simply misguided survival thoughts.

WHAT CAN YOU DO TO HELP?

In the first four parts of this book we showed you how you could help your child review a relationship and then discover and complete what was emotionally unfinished within the relationship. Using those same ideas, you would be able to review, discover, and complete what is emotionally incomplete between you and your former spouse.

The same technique is not as simple and clear in order for a child to complete their relationship with their parents' divorce. While it may not be obvious at first glance, the reason is that the child does not have a singular relationship to your marriage. The relationship is made up of at least three major components, possibly even more. There is the child's individual relationship with each parent, and then there is the child's relationship to the family itself. So it is quite a bit more complicated than most other relationships.

The reason it is not recommended that you try to help your children is that you cannot listen to them with a completely open heart. As they talk about your former spouse, you will automatically be inclined to "leave the moment."

So, if you can't help your children, who can?

LESLIE GETS THE LAST WORD, TOO

This section on divorce began with Leslie telling you the personal story of some of her reactions to her parents' divorce. Now as Leslie Landon Matthews, Ph.D., MFCC (Marriage, Family, and Child Counselor), she wants to tell you what she believes will be most helpful for your child.

I arrived at The Grief Recovery Institute just a few months after my dad's death. At the time, I was a practicing psychotherapist, specializing in working with children whose parents were either divorced or separated. John and Russell were instrumental in encouraging me to put in the time, energy, and effort to finish my doctoral studies and write my dissertation with a focus on children and grief. They are co-authors with me in the questionnaire that established the significant results that created my dissertation. Since then I have attained my doctorate in Marriage Family Therapy.

In all my years of study, I found precious few courses that would teach an evolving therapist what specific actions they could employ to help an adult or a child deal with the loss events in their lives. Even the class I had on Death and Dying, which is about the process of dying and not about grief and recovery, was an elective course.

While I was drawn to Grief Recovery by my father's death, it was in that first workshop that I became aware that I still felt incomplete about my parents' divorce. As I look back on that event, I realize that the workshop chronicled my newfound awareness that there was something missing from my education as a therapist. The ideas and actions that John and Russell were teaching made sense. What didn't make sense was that in all my training I had never been taught what they were doing. Encouraged by the personal emotional transformation that resulted from that workshop, I went back and attended The Grief Recovery Certification Program. I was able to see and learn that what had happened for me in the personal workshop was neither an accident nor a fluke. These guys knew what they were doing and could teach others how to do it, also.

I am a firm believer in therapy. And, I am a firm believer in the principles and actions of Grief Recovery. In my heart of hearts, I wish that all therapists could have the benefits of the Grief Recovery Certification training, as a vital adjunct to the other essential tools and skills they acquire on the way to their license requirements.

Many children's lives have been affected by loss experiences. For the most part, your children will benefit from the actions you take to complete your relationship with your former spouse. You automatically become a better guide and a better teacher. Your children may not need to take any formal actions to feel complete with your divorce. Like Jeff's daughter, they might just need a more complete parent who won't try to fix their feelings.

If your child is struggling in reaction to your divorce, then I would recommend that you look for a Certified Grief Recovery Specialist in your community. Some are therapists, some are not.

The important detail is that they each have the principles and actions of Grief Recovery to guide them, and, through their contact with The Grief Recovery Institute, they have direct access to John and Russell and the people they have trained.

One caution. As a parent, you have to be the judge of whether your child is in need of professional help. If so, take them to a licensed professional immediately. Even as I give you that warning, I would like you to be very careful about allowing your child to be placed on any psychopharmaceutical drugs. One of the dangers of drug intervention, for children and adults, is the covering up of the normal and natural responses to loss. Unless there are emergency circumstances, a nondrug approach may allow them to discover and complete the unfinished emotions attached to their reaction to your divorce.

I know that John and Russell join me in a most heartfelt desire that you and your children will be able to complete the pain, confusion, and sadness that may have accompanied your divorce. We cannot put marriages back together, and we cannot make the old dreams be real again. But we can help you all get your hearts back and then you can build new dreams. Good luck on your journey.

We are coming to the end of this book. Before we close up shop, we want to share with you some important information and a few personal success stories, which we know will be valuable. We know that you will find the stories both heartwarming and uplifting. More important, we hope they inspire you and give you courage to do more than just read this book. We hope you will feel encouraged to take the actions described throughout and become better equipped to help your children. They deserve it. So do you.

The stories were written by people who have had their own personal experiences with The Grief Recovery Institute, either directly in a personal workshop, or in one of the thousands of Grief Recovery Outreach Programs throughout the world. We had space for only a few stories, so we picked the ones that best represented a variety of life events that reveal the benefits of grief recovery work for children.

As an added bonus, we have reprinted the questionnaire that we wrote which led to Leslie's dissertation. You may find it fun and challenging to answer the questions, now that you have read the book.

CHAPTER 30

The "D" Word

Many parents are uncomfortable and even scared to use the words *dead* or *died* in their conversations with their children. We have observed parents spelling those words in front of their children, as if they were bad words. We recently heard a woman tell a friend that she did not want her children to see *The Lion King* because the daddy lion D-I-E-S, and that would be too traumatic for her kids. She actually spelled out the word so that her children would not understand what she was saying.

Before we go on, let us say that we do not want to be insensitive. Nor do we wish to dictate when your children should be taught about death. As a practical guide, your children's own awareness and response to the world they live in will usually help you recognize when they will be able to understand some simple and basic concepts about death.

The essential point is the same one that we have been making since the beginning of this book. That is, our task as parents and guardians is to prepare our children for the feelings they will have in response to the events that happen in and around their lives. The better the foundation we can give them, the better equipped they will be when losses occur.

ILLUSION OF PROTECTION

It is not uncommon for people to avoid a topic, in the belief certain words or ideas would negatively affect their children. Skirting sensitive areas may feel easier and even safer for the parent, but it may do a disservice to the children. Children are often more affected by the absence of truth than by the truth itself. One concern is that children, who are so tuned in to nonverbal communication, may have a sense that something is missing or not right in what they are being told. Avoiding a topic may create a short-term illusion of well-being, but putting off the truth may create larger problems later.

In the matters of death, dying, illness, divorce, and many other losses, the problem usually has less to do with those actual events than with the parents' lack of knowledge surrounding how to talk about those topics. We hope that what you have read in this book so far will strip away some of the fears you might have had and will allow you to talk more openly and honestly about loss. As you read through this book, you may have become aware that there are some events from your own past that may still be unresolved or incomplete. Completion of those long-standing issues will give you the freedom to communicate differently with your children.

SOLID AND CLEAR REFERENCE POINT

The danger in not giving your children a solid and clear understanding about reactions to death and other losses is that you leave their education on those topics to the whims and likely inaccuracies of others. In effect, you default that topic to the myths that we discussed in great detail in the opening chapters of this book. You must demonstrate and communicate openly about your emotional reaction to loss events. Silence or avoidance of the realities about loss creates more problems than it solves.

You do not need to become an expert. You do not need to know everything we know. You don't have to go back to school. You just have to begin to reconsider the ideas that you have been using to deal with loss. And then you need to start to put into practice the ideas you have read in this book. Go first. Tell the truth about yourself. You cannot hurt your children with the truth. You *can* hurt them with avoidance and lies, because those behaviors can create a separate grief issue—loss of trust.

SOMETIMES THE WORLD TRAVELS BACKWARD

Let us give you a very short history lesson about some aspects of a funeral that have changed over the years, and therefore are less helpful now than they were many years ago. Let's start in today's world and work backward. You may have attended what is called a visitation. (We are aware that not all religious and cultural customs are the same. In this instance, we are using one common practice as an example.) For those of you who don't know, a visitation is normally held at a funeral home and takes place within a few days after the death, before the funeral service and burial. The body of the deceased has been prepared and lies in state in an open coffin. Friends and family congregate to pay their respects, extend their condolences, and have some personal time to complete what is unfinished for them.

Funeral service professionals will have prepared the body for the viewing. Using a photograph, the cosmetologist will attempt to create a realistic image of the person who has died. The dressing and cosmetology are part of the services provided. Some of you may recall having been asked to pick out a dress or a suit in which a loved one would be buried. Often in today's world, picking out some clothes, or perhaps some jewelry, is the only participatory task performed by family members.

By contrast, in the late nineteenth century and for most of the first half of the twentieth, the visitation or viewing was in the parlor of the home. The cosmetology was the task of the family, most often the women and girls, who dressed the body and fixed the hair and makeup. During the process of preparing the body, the family was actively involved in what we call a relationship review. Naturally, as they made the preparations, they talked about the person they were preparing. The body itself acted as a stimulus to remind each person of their unique and individual relationship with the person who had died. Children learned to touch a dead body and to talk openly and honestly about the person and about death.

When funeral parlors came into existence, the death experience was removed from the home and handled in a more distant fashion. The term *parlor* was changed to *living room*, as popularized by *Better Homes Magazine*. Eventually, the word *parlor* more or less disappeared as it related to private homes.

One hundred years ago, it was probably reasonable to say that death was a part of life. When the process of preparation and the act of viewing were moved from the home to a a funeral home, something essential was lost: the element of *participation*. Implicit in the participation was a naturally occurring relationship review, which led to discoveries of incomplete events and emotions, and set up the possibility of completion.

Progress does not always result from change. Sometimes we go backward. The use of funeral parlors has had a major negative impact on both grief and recovery. It is hard to imagine parents using the phrase "D word" in front of children who have seen a body on view in the parlor of their own home.

In fairness to those of you who have been using this phrase, we are sure that you have done so with the belief that what you are doing is helpful and beneficial for your children. And you do not have to change if you disagree with us. But we want you to have some other information on which to base your decision.

TALKING ABOUT DEATH WITH YOUR CHILD

Back in chapter thirteen, in the section "The Death of a Pet," we suggested that some times are not appropriate for you to engage your child in an intellectual discussion about death. We said that we would give you some guidelines for the topic a bit later. Here they are:

There are several different ways that the topic of death can become important to your children. As with most things in life, regardless of age, awareness is always the beginning. Awareness usually comes from the observation of a change. It can be as simple as seeing a leaf green and alive on a tree one day, and brown and dead on the ground the next. Comparisons to objects in nature are helpful examples to assist you in communicating the basic meaning of death to a child.

Another opportunity to discuss this important topic can happen while taking a walk and seeing a dead animal. The sight of a dead squirrel or bird often gives a parent a timely opening to talk about death with their children. Your response to such a situation can have lifelong impact on them.

We could list all of the wrong things to say or do in that kind of situation. Instead, we'll discuss what is appropriate and effective. Imagine that you and your children are walking in the park, and your child sees a dead squirrel lying on the ground. This presents an ideal circumstance for you to explain what death means. For very young children, most of the ideas related to death will have to do with the obvious physical characteristics that they can see. You can help them notice that the animal is not moving. You can point out that the animal is not breathing. If there is an obvious visible injury, you can speculate as to how it might have happened. There may be an observation that the animal's eyes are open or closed.

One of the benefits of being this specific is that doing so removes any magical qualities and reduces the probability that your children will have inordinate fears about death. Left to their own devices and imaginations, children often arrive at incorrect conclusions. Accurate information helps them understand the world around them. It also helps them deal more effectively with emotionally affecting losses.

The things you do and say in reaction to seeing a dead animal are the cornerstones on which your children will establish their own beliefs about death. When you see the dead squirrel, you might say, "Oh, a dead squirrel, how sad." And in that sentence, both the factual truth and your emotional reaction to it are demonstrated by you to your child. The emotion of sadness is communicated as normal.

Throughout this book we have tried to avoid the feeling of a scientific text, because grief is emotional. In the next several pages, however, we will provide some more academic information to help you with a basic understanding of how young children deal with learning about death. It is essential that you acquire, demonstrate, and teach your children more effective ideas about death and grief and completion.

Sometimes adults don't realize how early in life a child has the capacity to mourn. By the age of eighteen months, children have the ability to preserve an image of a loved one, thus, they have the capacity to yearn for that person. When children are given truthful information and appropriate support concerning the circumstances of death, mourning in a manner similar to that of healthy adults is possible for them. Parents are responsible for how their children view death. When children can be shown how to view death in a realistic manner, they can attach the correct, natural emotions to the death.

CURIOSITY HELPS CHILDREN LEARN

Most children are extremely curious about death. Important concepts about death develop in a progressive manner. The issues of reversibility, universality, and causality are very important themes. Reversibility is the idea that the dead can return to life. Many children under the age of five do not understand that death is permanent and that all biological, cognitive, emotional, and physical functions have ceased for the dead. First, children learn that the dead cannot eat, sleep, or speak. Later they will begin to understand that the less obvious functions like thinking, feeling, and dreaming have also ceased.

Let us illustrate how the idea of reversibility affects different aged children. A father has died. Two months after Dad dies, the four-year-old son, the seven-year-old daughter, and their mother go on a trip. They are away for one month. Upon their return, as they pull into the driveway, the four-year-old spots his father's car in the garage and yells out with excitement, "Daddy's home! Daddy's back home!" The seven-year-old has a split-second feeling of excitement as well, then quickly realizes through her own higher level of understanding that her brother's comment is not true. She may sob, for she knows that Daddy is not home.

Leslie recalls that when a car killed her two dogs, she understood that they were dead. But even though she was eight years old, she still thought they could be hungry or tired. She remembers putting out food for them until she finally understood that they could no longer eat. Her parents did not understand at that time to sit down and explain to her what happens when living things die. She had returned to a youthful form of wishful thinking, which involves the wish for the deceased to return. Children will often hold on to habits and customs from the period before the death.

As you can see from Leslie's story, a little bit of correct information would have gone a long way toward helping her. Even if you are afraid of the topic of death, you might need to meet that fear head on so you can help your child. Understanding that death is not reversible is very important for your child.

Here is a statement that many of you will be relieved to hear and which will be helpful to everyone. A substantial amount of research indicates that:

> Giving children accurate information about the reality of death
> does not interfere with the development of religious
> or spiritual beliefs about heaven and afterlife.

We have separated and highlighted the above sentence to stress that the truthful communication about death does not inhibit or undermine religious or spiritual principles.

The concept of universality involves children's ultimate understanding that death is inescapable and that everyone must eventually die. Most studies indicate that by age six or seven children have a grasp of this concept. Younger children, however, are liable to ask, "Are you going to die, Mommy?" Or "Am I going to die, Mommy?" Those certainly can feel like awkward questions, but we believe that it is a good idea for you to be honest, clear, and direct in your statements about the issue. The timing of their understanding varies among children. Don't leave this to chance and guesswork. The better you educate them, the freer they will be to deal effectively with all aspects of life and death.

There can be major negative consequences when a parent does not answer those questions accurately. It is not unreasonable for our children to expect honest answers to their questions. Some time later, when they discover or realize that they have not been told the truth, they may have a hard time trusting the person who has misled them. That mistrust often translates into a larger sense of not trusting others, as well. We believe that if children can formulate the question, they deserve an honest answer. Even if they don't really understand what it means, they will always know they can trust that they are being told the truth.

The third important concept is causality. Causality deals with the physical-biological facts that lead to an understanding of the various causes of death. Children who understand the inevitability of death usually understand that there will have been a cause for the death. It is very important for a parent to listen very carefully to how their children believe the death occurred. It is not uncommon for children to engage in a form of magical thinking and to believe they are somehow responsible for someone's death. Sometimes if a grandparent has been stern or gruff with a child, the child will have a secret wish that the person will die. If the person dies within a short time after the child had that thought, the child may perceive him- or herself to have caused the death. It is very important that you stay tuned to the possibility of that kind of thinking so you can correct it.

Let us move forward for a moment and tell you that over the years we have seen an extension of that kind of thinking in adults, as well, who believe that their words, thoughts, or feelings caused someone to die. In every situation, that kind of thinking was traceable to a childhood event and beliefs that were never corrected with accurate information.

In summing up this section, we would caution you with this comment:

> Showing, telling, and teaching your children the truth can never hurt them—anything less runs the risk of lifelong negative consequences for them.

CHAPTER 31

Euphemisms + Metaphors = Confusion

For the next few moments this might seem like a high school English class. Sorry. Here are a few definitions from our worn but trusty *Webster's Ninth:*

Euphemism—the substitution of an agreeable or inoffensive expression for one that may offend or suggest something unpleasant

Metaphor—a figure of speech in which a word or phrase literally denoting one kind of object or idea is used in place of another to suggest a likeness or analogy between them

Analogy—1. inference that if two or more things agree with one another in some respects, they will probably agree in others; 2. resemblance in some particulars between things otherwise unlike: similarity

Simile—figure of speech comparing two unlike things that is often introduced by *like* or *as*—as in "cheeks like roses"; compare to metaphor

We don't understand it a whole lot better now than we did in English class years ago. But we want to make a very serious point here. Little children are very literal. They do not grasp the idea that words do not mean exactly what they say. Euphemisms and metaphors, in particular, are baffling to children, and when it comes to death and other losses, their use can have profoundly negative effects on them.

The classic example is the child standing in front of Grandma's casket being told, "Grandma's sleeping," whereupon that child is afraid to go to sleep for six months. As adults we understand that "sleep" is a metaphor for death; small children do not.

As we mentioned, children are very literal. As they begin to comprehend the world around them, they question everything. Almost any parent can recall the times when their little ones seemed to spend the entire day asking, "Why, Mommy?" or "Why, Daddy?" And, regardless of the answer, the child's follow-up question is, "But why, Mommy?" They are trying to make sense of their world. It is very important to their survival. Later they will add poetry, symbolism, and metaphor. Don't rush them and don't confuse them.

Another very tricky area is the use of religious and spiritual metaphors concerning heaven. We cannot calculate the number of calls we have received over the years from concerned parents whose children wanted to "go to heaven" to find Grandpa. There are even the rare but real examples of children having committed suicide to accomplish the goal of being in heaven with someone they loved who had died.

And on this same topic, when children hear phrases like "Daddy went to be with Jesus," there is a real danger that their literal interpretation can have unfortunate consequences. It is not uncommon for children to become angry with God or Jesus for taking away their loved one. We know of many people who never were able to complete the anger they felt as children, and who never established a loving relationship with God or any other religious principles. That is very sad, as they have been robbed of some of the potentially valuable benefits to be found in religious and spiritual sources.

Small children have a very hard time understanding things they cannot see or touch. Faced with an idea they cannot grasp, they will apply whatever they perceive about reality to that new idea. If Grandpa lives in the next town over, children know that we can get in the car and go visit Grandpa. If Grandpa has gone to heaven, that child's mind says, "Okay, let's get in the car and go visit Grandpa." Most people allude to heaven being up above. Small children take that very literally, also. We have been told many times of children looking out of airplane windows, trying to find Grandpa.

Earlier, when introducing some of the ideas regarding religious and spiritual concepts, we used the word *tricky*. This is tricky territory because, whatever you believe, you probably hold those beliefs strongly and without reservation. We are in favor of strong beliefs; we have some very strong beliefs about what we think are the best ways to help children deal with loss. The question here is, How do you communicate your beliefs about heaven to your children without confusing them?

For the most part, the answer lies in simply making sure that one key phrase is included in your communication with the little ones. In response to your children's questions, "Where's Grandma?" or "What happened to Grandma?" try this: "Grandma has died. We believe that after someone has died, he or she goes to heaven." The key phrase, which must be the *first* phrase, is "Grandma has died." It *must* be the first phrase; if you start with "Grandma went to heaven," your child will hear that and want to go there, and you may have a hard time saying, "But Grandma is dead, you can't go there now." Remember that little ones are completely literal.

Yes, you are liable to get a barrage of questions: What does died mean? What is heaven? Where is heaven? Who's at heaven? Can we go there? But you can answer each of those in turn. We spent some time earlier in this section giving you some guidance on explaining death to your children. We are not saying that this is easy. Just be honest, and remember that you must be simple and concrete.

When it comes to beliefs about heaven and afterlife, there are many different perceptions. Some people perceive heaven in a very literal sense with people in flowing white robes, surrounded by beautiful, idyllic scenery. Others have a more metaphorical image. Still others have no belief in heaven, in any form, and believe that death signals the end of a being altogether. It is up to you to decide how much of what you believe you communicate to your child. Whatever you decide, be very clear about telling your children about death.

We will end this section by restating something we said a few pages ago:

> Giving children accurate information about the reality of death does not interfere with the development of religious or spiritual beliefs about heaven and afterlife.

We hope that statement contributes to your ability to help your small children begin to understand death. Along with giving your children a clearer understanding of death, we encourage you to demonstrate the normal emotions that accompany the death of a loved one. Remember, your children look to you for guidance in all things. The factual reality of death is intellectual. The feelings attached to loss are emotional. The beliefs about what happens after this physical life ends are spiritual. We believe that all three are important.

CHAPTER 32

Four Weddings and a Funeral?

We have used the title of a popular film from a few years ago to introduce a very important topic. You may have noticed the question mark at the end of the heading. The question that we are asked more often than any other is: "Should I take my young child to the funeral?"

Without being facetious, we might ask, "Would you take your child to a wedding?" And you might be offended and suggest that weddings are very different. We would politely suggest that the difference is mainly in the kinds of feelings that are produced. Since this entire book has been dedicated to the idea that sad, painful, or negative feelings are a normal and natural part of life, then we would wonder why anyone would *not* want to take their children to a funeral.

The most important question is really whether or not your children are old enough to hang around, relatively quietly, while the adults do what they have to do. So, the same criteria you might apply to taking your children to a wedding would apply to a funeral. If they are old enough not to disrupt either of the proceedings, then they can go.

The bigger issue is that, in either case, you would have to educate your children if it is the first time they are going to either a wedding or a funeral. You must do some explaining *before* you go to the event. Let's look at a wedding. You would explain to your children what a wedding is. They may already know a little bit about that since their mommy and daddy are probably married. You would tell them that a wedding is where two people promise to love each other and help each other, and any of the other concepts about marriage that you wish to impart to them. You may include any religious or spiritual aspects that you associate with marriage. Then you explain the location, be it a church, a chapel, or an outdoor setting. You explain the kind of clothing generally called for, either formal or informal, depending on the style of wedding. And, most important, you explain that while the ceremony is going on they have to sit quietly and respectfully. That way, when they forget, you can remind them about that. And then you explain that a party or a dinner will be held afterward. Simply speaking, you would have prepared your children, in advance, for what to expect.

Now, that wasn't so hard, was it? The fact is, you've probably done that already. The question now shifts to the funeral. What would and should you do?

What if we say that you do exactly the same thing? Let's use the same guidelines for discussing the wedding to help explain the funeral to children.

First, you would tell your child that a funeral is something that happens after someone has died. It is a ceremony where we go to remember someone the way we knew him or her in life, so we can say good-bye. It is at this point, where you *might* share with your child whatever religious or spiritual beliefs your family has about what happens to the soul or spirit of the person who died. Notice that we highlighted the word *might*. Make your own decision about that, possibly guided by your children's interest or questions. You will explain where the event will take place—at a church or at the chapel of a funeral parlor, and later at a cemetery. You will tell your children what kind of clothing is called for to attend this ceremony. And, most important, you will tell them that they have to sit quietly and respectfully while the ceremony is taking place. You might mention that it is okay to cry—which also would have been okay at the wedding. You will explain, in advance, that a lot of people may be crying. They are sad because someone they love has died. You will have another chance to explain that "sad" is one of the normal feelings we have when people die. And afterward, we may go to a home or other place, where there may be food, and it might look like a party. The reason for that gathering is for people to remember and talk about the person who died. And, like the wedding, one of the other purposes is to witness the event and to share in the memories and the sadness.

Weddings and funerals are the same, only different. It's really easy if you don't scare yourself into thinking that it is somehow off-limits or out-of-bounds. If your children are just beginning to understand death, then the experience will be helpful for them. The degree to which you take the time and trouble to make a detailed explanation will give your children clear ideas of what to expect. You actually reduce their fears about death by telling them the truth. You cannot give your children better life preparation than to help them understand the emotional realities that relate to death in the same way that you educate them about the emotions generated by weddings and other positive events.

Because of the nature of our work, we have met thousands of people who say, "I don't go to funerals." Without exception, that statement and the fear attached to it can be traced to an unfortunate event that happened a long time ago. Most commonly, the individual was taken to a funeral as a child but was not told what to expect. The complaint we have heard most often is that there had been no advance explanation about "open caskets," and what it would be like to see a dead body for the first time. A simple explanation can have lifelong positive consequences for your children.

Not being adequately prepared for a funeral is just one example of an experience that can alter children's lives. Another very common experience, one with long-term negative impact, is when children were *not allowed* to attend the funeral of someone important to them. Twenty, thirty, and forty years later, they show up at our personal workshops still feeling tremendously incomplete, having been robbed of the potential for the elements of completion that a funeral can provide.

FORTY-FIVE YEARS LATER, BUT WHO'S COUNTING

Fred Miller is a dear friend of ours. From time to time we have asked him to review and comment on various sections of this book while we were writing it. As he was reviewing one of the first-person stories that appear in part six of this book, Fred had a very strong reaction. With his permission, we are going to tell you a little bit of Fred's story.

When Fred was ten years old, his maternal grandmother died. Young Fred did not have a very close relationship with this grandmother. He had not yet experienced the death of a family member, nor had he ever been to a funeral. He had no idea what to expect. His mom, most probably preoccupied with her own grief at the death of her mother, did not think to stop and explain to Fred what was going to happen at the funeral.

Fred went to the funeral with his mother. She did not force him to go. At the end of the service, the mourners filed past the open casket and paid their last respects. As Fred neared the casket, his mother told him to kiss his grandmother good-bye. Fred, out of respect for his mother's wishes, leaned forward to kiss his grandmother's face. As he moved, his hand touched his grandmother's cold, cold hand. He remembers an eerie feeling, which he almost can't describe, when his brain registered the sensation of his grandma's cold hand.

In a terrifying flash, he realized that his lips were about to touch Grandma's face. Since his body was already moving forward, he could not stop, and he kissed her. It was actually chilling to hear Fred recount that memory and to observe that, although forty-five years had gone by, Fred appeared to be reexperiencing the horrible sensations as if they were happening now.

That event has affected Fred's life in a variety of ways, not the least of which is the painful memory that reemerged as Fred read the sweet and wonderful story about the little girl and her NaNa who had died. Over his lifetime, Fred has been intolerant of others who are grieving, and even of his own grief. He once shunned a funeral director whom he happened to meet in a social situation. And he does not attend funerals. Fred can attend a viewing prior to a funeral, but he will not touch the body.

We are sad that Fred's story is true for him, but it really illustrates that there can be lifelong consequences when children are not given proper explanations and choices. While Fred's mother did not force him to kiss his grandma, when she told him to kiss her, it had the force of authority to a ten-year-old.

But the real reason we wanted to share this story with you is to suggest that with a little bit of explanation, four and a half decades of fear and discomfort could have been avoided.

Can you imagine the difference if Fred's mom, or anyone else, for that matter, had taken a few minutes and talked to Fred? It might have started with a couple of gentle questions, not an interrogation. Have you ever been to a funeral? Do you know what a funeral is? Have you ever seen or touched a dead body? This last question is very important. It can be a very strange sensation the first time a child—or an adult—sees or touches a dead body.

The brain can play some real tricks. You are looking at a human form, but since it is not animated, what you see can be confusing. Looking at a dead body is very different from looking at a sleeping person. Not just because of the stillness, but because of a certain kind of absence. There are many opinions about exactly what that absence represents. It is probably safe to say, without excluding any other ideas, that what is missing is the life force. As an adult, you can relate to what we have just said. You will have to find your own way to communicate that to your child.

Another eye-to-brain trick is that the dead body, which still looks like a person, does not feel the same as it did during life. We are used to feeling a certain level of heat and energy when we touch another living person. The kind of coldness, and the absence of energy coming off a dead body, can create a totally unfamiliar sensation. It can be terribly confusing as your brain reacts to a sensation it has never experienced. Confusing and scary.

If you don't know what to expect, touching a dead body can be an overwhelming experience. The event is further compounded by all of the emotions surrounding the relationship you had with the person who died.

A word to the wise: If your children are to go or be taken to a funeral, sit down with them first. Have a chat. Tell the truth about yourself. If you are a little scared or uncomfortable about having the chat, say that first. You may never have been to a funeral. You may not have ever touched a dead body. If those things are true, then drag out this book and read the last couple of pages out loud to your child. You can learn together.

Here are a few more things you can tell your child so he or she will have a better idea of why a funeral is so meaningful. Cultural and religious beliefs, as well as other considerations, dictate whether or not there is an open casket at a funeral or memorial service. One of the purposes of the funeral and the open casket (when applicable) is for people to see the body so they can be sure that what they heard is true. In a more technical way, we might say that it provides a visual confirmation of the information. This is very important. If you have ever wondered why so much time, effort, and expense is expended to find the bodies from those tragic airline crashes over the ocean, it is to provide the proof for the family that their loved one has died. That confirmation allows them to begin to discover and complete what may have been left emotionally unfinished at the time of the death.

One last thing: You may also explain to your children if they are going to a funeral that they may observe people talking to the dead body. The funeral service, if effective, will have helped those in attendance remember events in their relationship with the person who died. With those memories may come some awareness of things that were never said, along with the need to say some things again. Please help your children understand that it is healthy, normal, and helpful for them to say a few things at the casket. Of course, saying things out loud as well as touching the body are always optional. There should never be any force on either issue.

OUR COMPLETION WITH YOU

We have a little custom that we do at the end of each of our personal workshops and certification trainings. We ask the people in attendance if they have gotten something of value by participating in the event. When they say yes, we remind them that the value they received was the direct result of their courage in showing up and taking the new actions that we illustrated for them. We encourage them to give themselves a big pat on the back. We want to make sure that they realize that while we may have delivered the new information, they were the ones who were able to keep an open mind and acquire some new ideas, and with them some new freedom and new choices.

So, let's see if we can do the same thing here. If you believe that you have received something of value in what you have read in this book, then you deserve a big pat on the back for plowing through your old beliefs to arrive at some new ideas that will benefit your children. Our job is done. Now it is time for you to apply on a daily basis the things you have learned here. Be gentle with yourself, and be gentle with your children.

Before we say our good-bye, we want to enlist your support. If you believe that this book has invested you with some valuable tools that will enhance your children's lives, and if your financial circumstances allow, donate a copy to your public library, to your children's school library, to your church or synagogue, or to a social service organization. Encourage the teachers, therapists, and doctors in your community to get copies and make them available to their students and clients. Call your local funeral home and cemetery. Let them know that they must have a copy of this book in their community service lending library.

On behalf of your children, we thank you, we love you, and good-bye.

John, Russell, and Leslie

QUESTIONNAIRE

Earlier we promised to include the seventy-four questions that Leslie used in the research for her Ph.D. thesis. The results of her study showed clearly that young children process and complete loss far better if their parents have modeled correct grief recovery skills.

Leslie's research compared two groups between the ages of four and eight, each of whom had experienced the death of a close family member. In one group, the parent[s] or guardian of the child was familiar with The Grief Recovery Handbook and/or the Grief Recovery Personal Workshop or Outreach Program, and had a firsthand working knowledge of the principles and actions of Grief Recovery. In the other group, the parent[s] or guardian was not familiar with Grief Recovery. As you read the questions, you will notice that some of the questions reflect the fact that the adult filling out the questionnaire had experienced the same loss as the child.

1.   YES NO
When talking to my child about the death of his/her relative, I used the words "passed on": (i.e., "Johnny, I am sorry to tell you that your grandpa has passed on").

2.   YES NO
When talking to my child about the death of his/her relative, I used the word "died": (i.e., "Johnny, I am sorry to tell you that your grandpa has died").

3.   YES NO
Since the death occurred, I feel listless much of the time.

4.   YES NO
If I were walking through the park with my child and we came upon a dead squirrel, I would (or did) encourage my child to view the animal for as long as he/she wanted to.

5.   YES NO
If I were walking through the park with my child and we came

upon a dead squirrel, I would (or did) encourage my child to ask questions about the dead squirrel.

---

6.   **YES NO**
My child talks about positive events in his/her relationship with the person who died.

---

7.   **YES NO**
When talking to my child about death and grief, I told him/her to grieve alone.

---

8.   **YES NO**
When talking to my child about death and grief, I told him/her to express feelings freely.

---

9.   **YES NO**
When talking to my child about death and grief, I told him/her to cry if he/she wanted to.

---

10.   **YES NO**
When talking to my child about death and grief, I told him/her to keep busy so he/she wouldn't feel bad.

---

11.   **YES NO**
When talking to my child about death and grief, I told him/her to replace the loss (i.e., new friend, pet, etc.).

---

12.   **YES NO**
When talking to my child about death and grief, I told him/her not to express feelings outside of the family.

---

13.   **YES NO**
When talking to my child about death and grief, I told him/her to grieve for a few days and then move on with his/her life.

---

14.   **YES NO**
When talking to my child about death and grief, I told him/her to be strong for others.

---

15.   **YES NO**
I expressed my feelings about the death in front of my child.

---

16.   **YES NO**
My child cried about the death of the relative in front of me.

---

17.   **YES NO**
I am preoccupied with the death of my relative much of the time.

18. YES NO

I encouraged my child to talk about the future events that will no longer be shared with the relative who died.

19. YES NO

My beliefs about death and grief will influence my child.

20. YES NO

I am still experiencing change in my sleeping pattern.

21. YES NO

I talked about positive memories and events in my relationship with the person who died.

22. YES NO

I would (or did) explain to my child what he/she might expect from seeing or touching the dead body (i.e., coldness of skin; stiffness of arms and legs; different skin color, facial expression, etc.).

23. YES NO

Since the death occurred, I have noticed my child has less interest in social activities and friends.

24. YES NO

I would (or did) explain funerals to my child.

25. YES NO

Since the death, I have put away any pictures that were around our house of the relative who died.

26. YES NO

I would (or did) talk to my child about what happens to the body after the death occurs.

27. YES NO

After explaining what funerals are about, I would (or did) give my child the choice of going to the funeral service.

28. YES NO

I believe that children should be notified of the relative's death as soon as possible.

29. YES NO

I encouraged my child to grieve alone after he/she found out about the death.

30.   **YES NO**
Since the death occurred, I have noticed my child acts listless much of the time.

---

31.   **YES NO**
I was able to discuss my feelings about the death of my relative with someone close to me.

---

32.   **YES NO**
I encouraged my child to talk about the relationship that he/she had with the relative who died.

---

33.   **YES NO**
Since the death occurred, my child has cried openly.

---

34.   **YES NO**
I encouraged my child to talk about the positive memories and events in his/her relationship with the person who died.

---

35.   **YES NO**
I would (or did) encourage my child to see the body if he/she wanted to.

---

36.   **YES NO**
I would (or did) encourage my child to pick out or accept a personal item that belonged to the relative who died.

---

37.   **YES NO**
I would (or did) encourage my child to participate in the funeral ceremony (i.e., writing a note, drawing a picture, singing a favorite song, etc.).

---

38.   **YES NO**
When the death occurred, I told my child how I felt.

---

39.   **YES NO**
I used my memories of my own loss experiences to explain grief to my child.

---

40.   **YES NO**
Since the death occurred, I have less interest in social activities and friends.

---

41.   **YES NO**
I talk about the negative memories and events in my relationship with the person who died.

42. YES NO

I told my child that I knew how he/she felt.

---

43. YES NO

I prefer to grieve alone.

---

44. YES NO

I encourage my child to cry.

---

45. YES NO

I told my child that he/she might experience trouble paying attention following the death of his/her relative.

---

46. YES NO

I told my child that death can provoke some questions about life (i.e., How long do people live? Does everybody die? Why do we die?).

---

47. YES NO

I encouraged my child to talk about his/her relative whenever he/she wanted to.

---

48. YES NO

When our relative died, my child expressed his/her feelings about the death.

---

49. YES NO

I talked about the relationship that I had with the relative who died.

---

50. YES NO

I encouraged my child to be strong for the rest of the family.

---

51. YES NO

I believe that children should be spared the knowledge of a relative's death for as long as possible.

---

52. YES NO

My child talks about the relative who died whenever he/she wants to.

---

53. YES NO

My child talks about future events that will no longer be shared with the relative who died.

---

54. YES NO

I am able to cry about the death of our relative in front of my child.

---

55. YES NO

When my child asked me questions about the death, I answered the questions accurately, to the best of my knowledge.

---

56.   **YES NO**
When my child asked me questions about the death, I changed the subject to happier topics.

57.   **YES NO**
When my child asked me questions about the death, I provided answers that I thought the child would like to hear.

58.   **YES NO**
When my child asked me questions about the death, I told him/her that I did not want to talk about it.

59.   **YES NO**
When my child asked me questions about the death, I told him/her that it would be just a matter of time before he/she felt better.

60.   **YES NO**
After the death occurred, I encouraged my child to talk about the negative events in the relationship he/she had with the person who died.

61.   **YES NO**
My child expresses his/her feelings in front of me.

62.   **YES NO**
If or when a pet dies, it is best to replace the loss immediately so a child will not feel sad.

63.   **YES NO**
I encourage my child to keep busy so he/she won't feel sad.

64.   **YES NO**
I want to stay strong for my family.

65.   **YES NO**
I do not like to cry in front of my family.

66.   **YES NO**
I would (or did) encourage my child to ask questions about what happens to the body after death.

67.   **YES NO**
When the death occurred, I encouraged a time for sharing questions and feelings with my child.

68.   **YES NO**
Since the death occurred, I have had some difficulty concentrating.

69.   **YES NO**
I encouraged (or would encourage) my child to go to the funeral or viewing of the body if he/she wanted to.

70.   **YES NO**
I demonstrate my emotions in front of my child (i.e., crying).

71.   **YES NO**
I told my child that it is common to have dreams about a loved one who has died.

72.   **YES NO**
Since the death occurred, I keep busy so I won't feel sad.

73.   **YES NO**
I talk about my relative who died freely and openly in front of my child.

74.   **YES NO**
I talk about future events that will no longer be shared with the relative who died.

Acknowledgments

From John:

As I look back over these past many years, I see clearly now that this has been a spiritual journey for me. One that has been filled with miracles. In the beginning I didn't see them as clearly as I do now. In fact, I see them so clearly now that I expect them to happen every day. And they do!

They appear in the eyes of people who bring in broken hearts for us to mend. We provide information that they don't have, and then we wrap a blanket of emotional safety around them so they can use this new information to take new actions. The results of those new actions include emotional freedom and a new sense of joy and purpose in life. These loving and courageous people communicate this transformation to us with the smiles on their faces and the look in their eyes.

To be privileged to watch this renewal in the eyes of a child is the greatest miracle of all. To see hope return to a child fills my eyes with tears, and I refuse to hide them as I once did. These miracles do not come from us; they come through us.

I want to thank, in some small way, Russell and Leslie. They have worked tirelessly to complete this work. Together, we hope and pray that this book will be an essential book for all parents. There is no greater gift that parents can give children than the ability to heal their hearts when they get broken, as they will many times during their lives.

I want to acknowledge and thank my two children, Allison (twenty-five) and Cole (nineteen). They keep getting older and wiser. I just keep loving them.

Lastly, I want to thank my wife, Jess Walton. Even as I write those words, tears come to my eyes. She has been everything to me for twenty years. (This year, we celebrate our twentieth wedding anniversary). When we met, it would be fair to say that I had some rough edges. She has smoothed many of them. It would be fair to say that I had few social graces. She has given me what grace I now possess. During my darkest days, she has been the one to shine her light on me. Her light is powerful and bright.

She is a famous and beautiful actress. She has been on the number one–rated daytime television show for many years. She has won many awards and even has two Emmy's on the shelf. With these few words, I want her to know that I have no awards to give. I have only my integrity, my loyalty, my word, and my love. And, in the end, I guess those are the things that really count.

I LOVE YOU—that's "computerese" for *I'm shouting*.

From Russell:

It is Sunday morning, and I am sitting at the desk at which I have sat for many years. It is from this spot that I have talked with more than fifty thousand grieving people, each of whom called here because his or her heart had been broken by a singular loss, or by the cumulative losses of a lifetime. For me, that their hearts had been broken, and that they did not know what to do to help themselves or their children deal with the emotional pain they were experiencing, became the central motivation for this book.

Working with grieving people has given us a vast storehouse of experience and knowledge. Yet month after month, year after year, each call had to start from scratch. The callers didn't know what we know, they knew only the things they had observed or had been taught as far back as they could remember. The common denominator that connected all of the callers was the information or misinformation they brought to their own loss experience. Without exception, the ideas they were using were ideas that they had learned when they were very young. They had practiced those ideas over and over and had never questioned them. Sadly, incorrect information, no matter how often it is used, never produces helpful results.

Anyone in the mental-health field could probably tell you that people who call for help sometimes argue with the advice they receive. Seems funny, doesn't it? When you call the plumber or electrician for help, and you don't have a clue about those trades, you usually don't argue with them (of course, you may argue when the bill comes). In truth, of the fifty thousand callers to whom I have talked, I have "argued" with probably at least half of them, because what *they* believe and what *we* teach are almost always exact opposites. And a few of them have gotten upset with me . . . but very few, I think five, altogether (but who's counting?).

For many of the callers, there was no argument. For them, it was more of a confirmation about the experience they were having. For them, it was a validation of their sense that people were avoiding them or not mentioning the topic of their loss—and that their sense was real, they were not being paranoid. For many of them, just to hear the simple statement from us, that what they were experiencing was "normal and natural," was enough.

So this acknowledgment is to thank the 49,995 grievers who allowed me to give them different ideas so they could have the hope and possibility of getting different results, and to thank them for being the underlying force that dictates the need for this book.

Along with John and Leslie, I firmly believe that if your children learn better ideas for dealing with loss, they may never have to call here, or come to one of our workshops, because they would already have the correct ideas to deal with losses of any kind.

And now for something completely personal. Alice Borden and I have been together for thirteen years. My first indirect contact with Alice was on her telephone answering machine, after having been given her number by a mutual friend. Being nervous, calling a stranger, I bumbled out a few words to explain who I was. And then I said, "I feel like a teenager, making this call." She maintains that it was the honesty of that comment that encouraged her to call me back.

All these years later, I still feel like a teenager with Alice. It's a feeling I hope I never lose. Thanks, honey.

From Leslie:

I would first like to dedicate this book to my husband, Brian, and thank him for all the support he has given me. You are my best friend, my soul mate, and the love of my life. To my children, Rachel Lynn, Justin Michael, and Catherine Michelle, thank you for your patience and understanding while Mommy worked on this book. You are all my "greatest joy" in life, and my heart is filled every day by the love we have for one another. I would also like to thank my mother, Lynn, and my family for their support, encouragement, and, most of all, their love.

I would like to acknowledge and thank Russell and John for inviting me along on this wonderful adventure, and for giving me invaluable information that I hope I can use to help others.

And to my dad, Michael, who will reside in my heart forever.

From Eric Cline
(Director, Grief Recovery Educational Society, Canada)

It's hard for me to believe that The Grief Recovery Institute of Canada just celebrated its tenth anniversary. Although I publicly thanked John and Russell in *The Grief Recovery Handbook,* I want to restate my appreciation for the time, love, and support they have given to this organization. With their tireless help, The Grief Recovery Institute of Canada and The Grief Recovery Educational Society (a nonprofit charitable organization) have become valued resources throughout the country for both grievers and the care-giving community. Together we have introduced a process of recovery that actually enhances life in an environment that can so easily limit our ability to participate in life. I know there are a great many people who join me in saying, "Thank you, and we love you."

And my special thanks to Jess and Alice for your support and understanding when I've dragged John and Russell so far from home or called in the middle of the night seeking advice. To Celina, my special little friend, for understanding when I was away so much and keeping me up to speed by phone, and for lovingly welcoming so many other children and families into our home. You are in my heart and my prayers. I love you! To my parents, Richard and Jean, sisters, Elizabeth and Cathy, and their families, for your love, support, and for helping out when I couldn't be in three places at once. To everyone who has helped with charity events, creating grief recovery awareness, and to all our grief recovery certified people who have invested their lives, their love, and their time delivering Grief Recovery Outreach Programs.

From All of Us:

To our HarperCollins family and especially Trena Keating, Gail Winston, and Christine Walsh.

A BOOK ABOUT LOSS EXPERIENCES A LOSS

For many years, our personal champion has been our editor, Trena Keating. When Trena sent us the edited copy of the original manuscript of this book, she sent with it a letter. One paragraph of the letter was so powerful that it brought tears to our eyes. Her comments showed us how clearly Trena understood what we were trying to say, and reinforced our belief that this book could truly help people help their children.

And then, as often happens in real life, there was a major change in our family. Trena moved on to a different job at a new company, and for a moment it was as if this book had lost one of its parents. Since this book is about dealing with loss, we had to practice the principles that we teach. So we had to grieve and complete our relationship with Trena so that we could begin a new one with our new editor, Gail Winston. The miracle is that Gail has adopted this book as if it were her own child, and has helped nurture it so it can get from us to you and to your children.

During the transition, a most important person, Christine Walsh, assistant to Trena and now to Gail, stepped in and bridged the gap, helping us to feel safe that our baby was in good hands. Thank you, Christine.

We sincerely hope that you and your children derive benefit from this book. When you do, please tip your ❣ to Trena, Gail, and Christine.

THE GRIEF RECOVERY INSTITUTE
SERVICES AND PROGRAMS

Recognized as the foremost authority on recovery from significant emotional loss, The Grief Recovery Institute offers an ever-expanding range of programs for individuals and organizations throughout the world.

Programs currently available include:

The Grief ❣ Recovery® Certification Program: This four-and-a-half-day event is the most powerful and comprehensive training program of its kind offered anywhere in the world. Graduates return to their respective communities as Grief ❣ Recovery® Specialists, fully prepared to establish and maintain Grief ❣ Recovery® Outreach Programs and Community Education Forums. Thousands of such programs now exist in communities throughout the world. Certified alumni have constant access to and support from The Grief Recovery Institute. Certification training is conducted regularly in Los Angeles and monthly in select cities in the United States and Canada.

The Grief ❣ Recovery® Personal Workshop: The highly intensive, three-day, recovery-focused experiential events are limited in size (ten to fourteen), to allow each participant as much personal attention as required. These life-altering workshops are attended by individuals, couples, and families who are struggling with a devastating loss or an accumulation of losses. Personal workshops are conducted regularly in Los Angeles and monthly in select cities in the United States and Canada.

The Grief ♥Recovery® Outreach Program: The outreach programs are moderated by Certified Grief ♥Recovery® Specialists. The twelve-week program is recovery focused and follows a format created by the institute, based on the principles and actions outlined in *The Grief Recovery Handbook*. We do not call it a "support group," specifically to avoid the possibility of it becoming a group that accidentally supports pain instead of recovery. Nor is it a self-help group, because it has a leader-facilitator and a clear format of actions. And lastly, the outreach program is not therapy. It is most accurate to call it an educational or reeducational experience with positive benefits to all who participate.

The Grief Recovery Handbook: The Action Program for Moving Beyond Death, Divorce, and Other Losses, available in libraries and bookstores, is the ultimate resource for anyone struggling with the aftermath of any kind of loss.

At long last, *The Grief Recovery Handbook* and the outreach program have been translated into Spanish and will be available in the spring of 2001.

Long overdue and now a reality are:

The Grief ♥Recovery® Children's Workshop for Parents (Guardians, Teachers, Clergy, and Others)

The Grief ♥Recovery® Children's Program for Health Care Professionals

The Grief Recovery Institute Educational Foundation, Inc. (a nonprofit, 501[c][3] corporation) provides speakers for educational presentations for school districts, municipal and service organizations, and radio and television appearances, as time permits.

For information on any of our programs, please contact us.

In the United States:
    The Grief Recovery Institute
    P.O. Box 56223
    Sherman Oaks, CA 91403
    (888) 773-2683

In Canada:
    The Grief Recovery Institute of Canada
    188 Charlotteville Road W. 1/4 Line
    R.R. #1
    St. Williams, Ontario
    Canada   NOE IPO
    (519) 586-8825

We also hope that you will visit us on the World Wide Web. Our home page is located at:

www.grief.net

You may also e-mail us at: support@grief.net